Ivan Kalaš Roland T. Mittermeir (Eds.)

Informatics
in Schools

Contributing to 21st Century Education

5th International Conference on Informatics in Schools:
Situation, Evolution and Perspectives, ISSEP 2011
Bratislava, Slovakia, October 26-29, 2011
Proceedings

 Springer

Volume Editors

Ivan Kalaš
Department of Informatics Education
Comenius University
842 48, Bratislava, Slovakia
E-mail: kalas@fmph.uniba.sk

Roland T. Mittermeir
Institut für Informatik-Systeme
Universität Klagenfurt
9020 Klagenfurt, Austria
E-mail: roland@isys.uni-klu.ac.at

ISSN 0302-9743   e-ISSN 1611-3349
ISBN 978-3-642-24721-7   e-ISBN 978-3-642-24722-4
DOI 10.1007/978-3-642-24722-4
Springer Heidelberg Dordrecht London New York

Library of Congress Control Number: 2011938015

CR Subject Classification (1998): K.3, J.1, K.8, H.5.2, D.1, D.3

LNCS Sublibrary: SL 1 – Theoretical Computer Science and General Issues

Typesetting: Camera-ready by author, data conversion by Scientific Publishing Services, Chennai, India

Printed on acid-free paper

Springer is part of Springer Science+Business Media (www.springer.com)

Preface

The International conference on Informatics in Schools: Situation, Evolution and Perspectives, hosted by the Comenius University in Bratislava, was the 5th in the series of ISSEP conferences.

The series started in 2005 under the title "International Conference on Informatics in Secondary Schools: Evolution and Perspective" in Klagenfurt, Austria on the occasion of the 20th anniversary of compulsory formal informatics education in Austrian secondary schools (gymnasia). A substantial aim was to bring local teachers into contact with developments and ideas forming the basis of informatics didactics in other countries. Consequently, the proceeding were split into two parts. One appeared in the Lecture Notes in Computer Science series of Springer Verlag [LNCS 3422], the other one, published by Ueberreuter, a local publishing house, gave both, Austrian teachers and international visitors, a forum for presenting "Innovative Concepts for Teaching Informatics".

The concept of split proceedings as well as the idea of bringing local teachers in contact with the international community has been preserved over the years. The focus of topics shifted slightly. At the second ISSEP, held in 2006 in Vilnius, Lithuania, the overall theme was "Informatics Education – The Bridge between Using and Understanding Computers". Thus, the spectrum addressed ranged from programming and algorithmics via ICT-education to e-learning [LNCS 4226]. In the accompanying volume, "Information Technologies at School", published by TEV Pulishing House, e-learning, ICT, and Informatics competition are widely described. Corresponding to the theme "Informatics Education – Supporting Computational Thinking" [LNCS 5090] the third ISSEP, held 2008 in Toruń, Poland, shifted the focus away from ICT and e-learning. One section of the accompanying proceedings, published by Nicolaus Copernicus University, Toruń, had in contrast a section linking informatics with mathematics instruction. This trend continued in both proceedings of the fourth ISSEP, January 2010 in Zürich. It focussed on "Teaching Fundamental Concepts of Informatics" [LNCS 5941].

For the fifth ISSEP, held in fall 2010 in Bratislava, Slovakia, it was decided to broaden the scope by including aspects of informatics or informatics-related education also at lower grades. This led to a reinterpretation of the ISSEP acronym to "International conference on Informatics in Schools: Situation, Evolution and Perspectives". There were two reasons for considering the spectrum of formal IT-related education over all age groups, i.e., including primary school and for some countries even aspects of IT-related motivational activities in Kindergarten [3].

Firstly, this expansion of scope is a consequence of trends to be observed in various countries, where some aspects of informatics-related education have been shifted from the secondary to the primary level. This shift is a consequence of changes in educational policy as well as of the fact that informal (but incomplete) peer-group instruction and factual use of modern information technology (from cell-phones to home-computers), have penetrated the life of even very young kids. Since this early educational confrontation of children with information technology has consequences on secondary level curricula, extension of the scope became a necessity. A second reason for extending the scope of the ISSEP-conferences is the fuzziness surrounding the definition of "secondary education" in various countries. While formal education starts in most countries at the children's age of 6 or 5, the duration of primary education varies. A range of 4 to 6 grades is rather common. Due to different national stratification of the educational system, the variance in duration and age-groups embraced by secondary education is even broader [2]. Consequently, depending on nation and gender, the shift from primary to secondary education falls into the critical period of beginning adolescence. This is a good reason not to focus on organizational structures of national school systems but consider the whole spectrum of school and ask authors to be specific about the age group or grade they are referring to in their contributions.

Out of 69 submissions from 20 countries the Program Committee selected 20 papers from authors of 12 countries for inclusion in this volume[1]. Each paper was reviewed by at least three members of the Program Committee in an electronic reviewing process. The papers in this volume have been arranged into the topical groups briefly described below.

This volume is opened by three papers representing the *Spectrum of Options* to be considered in *Informatics Education*. Pavel Boytchev shows the wealth of a creative learning process during which a student (apparently highly motivated) acquired programming knowledge on her own in an investigative process with minimal guidance. Valentina Dagienė leads readers into an opposite corner of the educational spectrum by addressing a host of informatics-related topics relevant for contemporary learners. The breadth of topics addressed is due to various recommendations of international bodies that are mainly concerned with economic and political issues. Therefore, they are also concerned with guidance for (inter-)national education policy. The ensuing paper by Juraj Hromkovič and Björn Steffen might be seen as a counter position they share with a substantial part of the scientific community involved with informatics education. Thus, it defines a further corner point specific for informatics instruction: the need to present pupils the core-concepts of informatics. This aims to present informatics as a science contributing to our contemporary society. Obviously, these concepts are of a more fundamental nature than short-lived application-oriented skills, part of society is asking school to teach.

Considering these papers individually, one or the other participant of the conference might put forward valid counter-arguments resulting from a particular position of teaching informatics in school. However, seen in conjunction, the arguments voiced in those papers span an area and it is important that educators

---

[1] All of the other papers presented at ISSEP 2011 are included in the accompanying conference proceedings [4].

perceive this area's full breadth. A school-class is not an amorphous aggregate of humans pertaining to a given age group. It is a collection of individuals. Teachers have to bridge the gap between the prescription of the curriculum to be followed for all pupils and the particular educational needs of each individual student, be these due to special abilities or special deficiencies, as addressed in the paper from the Japanese group on teaching handicapped students.

Out of this group of papers one can also sense a terminological tension between the terms "informatics" and "computer science". Most countries use just one of these terms when referring to informatics or IT-related instruction. In other countries the distinction between using information technology (IT- or ICT-related instruction) and providing insight into the conceptual and technological basis of modern information technology is made explicit in the curriculum. Authors have been asked to clarify whether they are referring to instruction concerning basic conceptual elements of the discipline or to instruction concerning skills on using information technology in order to act as a versatile citizen in a modern and well-developed economy.

The second section of the proceedings contains three papers grouped under the heading *National Perspectives*. The first paper, with Maria Carla Calzarossa as leading author of a team consisting of four scientists from four Italian universities, reports on the way ICT-concepts are taught in schools belonging to eight regions spread over Italy. It is followed by a paper from Peter Micheuz describing the results obtained by a working group aiming to align ICT-instruction with engraining aspects of computational thinking for the pupils belonging to the age-group from 10 to 14. Maciej Sysło's paper on Outreach to Prospective informatics Students establishes the link to the next section. After describing recent developments in the curriculum of Informatics for Polish schools of various levels, it provides an overview of the broad spectrum of Polish outreach activities.

The section *Outreach Programs* commences with an overview paper written by an international team of five authors describing and comparing outreach programs they once proposed and that are now internationally well recognized. The spectrum ranges from Computer Science Unplugged (Tim Bell) via cs4fn (Paul Curzon) and CS Inside (Quintin Cutts) to Bebras (Valentina Dagienė), and the Israeli approach of supporting pupils specializing in CS in a comprehensive venture by bringing them into direct contact with well-known academics and having them participate in an industrial software engineering project (Bruria Haberman). The second paper in this section focusses on CS-Unplugged in a non-traditional manner. Hiroki Manabe and his co-authors report on lessons they conducted with handicapped pupils in Japanese schools. Due to the physical constraints of these persons, applying CS-Unplugged as originally proposed by Bell, Witten, and Fellows [1] is unsuitable or even impossible. Hence, the authors provide these pupils with a computer interface that allows them to overcome senso-motoric or other physical deficiencies in a simulated environment. Ernestine Bischof and Barbara Sabitzer, concluding this section, describe on an exemplary level a project where, similarly to CS-Unplugged, core computer science topics have been framed into contexts even young pupils can relate to. The paper mainly reports key results from the evaluation of the interventions done in schools. Among those results was the observation that marked gender differences became noticeable only with older age groups. They were not noticeable in primary schools. Attentiveness to these interventions in primary school was highest among the age groups studied. Hence, arguments suggesting that core-concepts of informatics have to be reserved to age groups past 12 years, or CS as a technical subject is more attractive to boys than to girls, have to be considered with scepticism or – considering also other papers addressing this topic – have to be dismissed. Pointing at deficiencies in teachers' capabilities to adopt this or a similar approach on their own, links to the next section.

Teacher Education is apparently a key side condition for good informatics education in any country. Since a substantial portion of active teachers got their teaching degree before informatics had been defined as regular subject in school, in-service teacher education plays a critical role in informatics. Noa Ragonis and Anat Oster-Levinz focus on a practicum program of students enrolled in a pre-service teacher education degree program. A specific focus of this paper is on the multi-facetted evaluation program these student-teachers are subjected to. The other paper on teacher education was contributed by Daniela Bezáková and Michal Winczer. It explains the author's approach to teaching informatics teachers concepts of theoretical informatics. The approach is directed to pre-service as well as to in-service teachers. It particularly addresses issues resulting from the students' limited background in mathematics. The significance of this approach is due to the fact that especially in the generation of teachers perceiving informatics education only from an ICT-perspective, theoretical informatics plays a secondary role. They consider the concepts it provides too abstract and not sufficiently applicable in the pupils' daily life. One has to question though, whether such arguments are not mainly put forward as self-defense to cover up didactical deficiencies. A series of counter-arguments to such a short-sighted line of arguments are raised in the paper by Hromkovič and Steffen appearing in the introductory section "Spectrum of Options".

The section *Informatics in Primary Schools* is opened by Andrej Blaho and Ľubomír Salanci's report on the principles pursued when developing an informatics curriculum for Slovakia. In Fig. 1 of their paper, the authors give a succinct overview of the various forces and needs to be considered when establishing an informatics curriculum for (primary) school. Before presenting the proposed approach they discuss why compromises have to be found between purist proposals. The paper closes with evaluative reflections on an initial implementation of the recommendations. Giovanni Serafini describes outreach activities of a team from ETH Zürich in order to teach pupils of (remote) primary schools computational thinking. The age group addressed consists of children between 8 and 13 years old. The programming language used has been Logo. But the aim of the approach was certainly not teaching some programming language but rather using Logo only as medium for kindling computational thinking. Interventions in two schools are described in detail. Gerald Futschek and Julia Moschitz also aim at instilling computational thinking with very young kids. Instead of a computer-executable programming language they use cards for representing commands needed to load or unload colored wooden blocks into a toy train. Kids are challenged by arranging the cards in such a way that a robot can correctly load the train. Evidently, the arrangement of cards results in a "program" controlling the movement and work of this assumed robot. Later, the arrangement of cards might even be simulated in an environment like Scratch.

The section *Advanced Concepts of Informatics in Schools* is opened by a paper from David Ginat, Eyal Shifroni, and Eti Manashe on difficulties students have with the transfer of previously acquired knowledge when solving new programming tasks. The paper addresses these problems on the basis of five particular example-problems and the solutions as well as difficulties kids had in solving them due to specific transfer problems. Viera Krňanová Proulx focusses in her paper on "Program by Design". A systematic approach to problem solving following a well-structured design process is advocated. The approach is explicated by the description of design recipes for functions and methods, data definitions, and abstractions. The functional approach is extended for issues arising with object-oriented programming. The paper by Lucia Keller and her co-authors from ETH-Zürich presents a course on classical cryptology, offered by the Informatics-Didactics group of ETH to Swiss schools. Thus, this paper could also be seen as a further extension of the "Outreach"-section. The description of the approach is detailed enough for other teachers to follow it. However, in doing so, one must not ignore an aspect easily overlooked on first reading. The innovative idea is not to teach cryptology. It is rather to briefly present pupils some cryptosystem and have them break enciphered messages. This approach is didactically more elaborate than mere frontal presentation. However, the motivational effect and creativity stimulation obtained is apparently substantially higher.

The volume closes with a section on *Competitions and Exams*. The two papers on competitions might be seen as further extension of outreach programs, but all three papers have the proper composition of questions posed or the organization of the competition in their focus. Hence, they have been grouped into a common section. Monika Tomcsányiová and Peter Tomcsányi extend the widely known Bebras contest by proposing a "Little Beaver" contest for children aged between 8 and 9 years. After briefly describing Bebras, the authors contrast their approach to the contest targeted at pupils of 10 years upwards. They mention not only limits in the knowledge domain and abstraction capabilities but also senso-motoric and other general constraints to be considered when addressing a very young age stratum. On this basis, a pilot run was developed. The results obtained are described. An approach of developing a contest for pupils following the approach of the Kangaroo-contest in mathematics is reported in the paper by Violetta Lonati and her co-authors from the Università degli Studi di Milano. Their approach is explicitly defined as an outreach program for all pupils of a class. It should not depend on some specific prior training. The contest consists of two rounds asking questions of different complexity. A school-internal qualifying round is followed by a nationwide final round. The test examples of the final round are (re-)used in in-service teacher education. The section closes with a paper on criteria for writing examinations. Haim Averbuch, Tamar Benaya, and Ela Zur report on an analysis of school-internal exams used to prepare students for the final matriculation exams in computer science conducted in Israel on the national level. Even if the paper and the study it reports upon is motivated by specifics of a top-level national exam, the criteria established by the authors are worth considering when designing any sort of comprehensive exams in informatics/CS.

Before closing, I should mention that a conference like this is not possible without the support of many individuals and organizations. Hence, I would like to thank particularly the General Chairman, Prof. Ivan Kalaš, and all members of the Program Committee as well as all additional reviewers for ensuring the quality of the proceedings. Carrol Sperry deserves special mention for helping some authors to improve the linguistic aspects of their respective paper. Special thanks go also to the Organizing Committee. I also have to be grateful to Annette Lippitsch for editorial support in copy-editing the papers contained in these proceedings.

The conference was made possible by the support of several sponsors. Among them I would like to single out the Slovak Society for Computer Science, Slovenská informatická spoločnost', for supporting student-participants and the Austrian Ministry for Education, Art, and Culture, Bundesministerium für Unterricht, Kunst und Kultur, for financial support of attending Austrian teachers and the production of the conference proceedings. Finally, we are grateful to the Comenius University, Bratislava, for hosting the conference.

References

1. Bell, T., Whitten, I.H., Fellows, M: Computer Science Unplugged: An enrichment and extension programme for primary aged-children, (2005),
   http://csunplugged.org/.
2. Hubwieser P. et al.: Informatics in Secondary Education, Report on ITiCSE'11, Working Group 2, to appear.
3. Kalaš, I.: Recognizing the potential of ICT in early childhood education. UNESCO Institute of Information Technologies in Education, Moscow, 2010. 148 p. ISBN 978-5-905175-03-9. Available also on-line at iite.unesco.org/publications/3214673/
4. Kalaš, I. (ed.): Proceeding of the 5th ISSEP – The International Conference on Informatics in Schools: Situation, Evolution and Perspectives. Bratislava, Comenius University, 2011. ISBN 978-80-89186-90-7.

August 2011   Roland Mittermeir

Conference Organization

Conference Chair

Kalaš, Ivan	Comenius University, Bratislava, Slovak Republic

Program Committee

Kalaš, Ivan (Chair)	Comenius University, Bratislava, Slovak Republic
Dagienė, Valentina	Vilnius University, Lithuania
Ginat, David	Tel-Aviv University, Israel
Futschek, Gerald	Technische Universität Wien, Austria
Hromkovič, Juraj	ETH Zürich, Switzerland
Micheuz, Peter	Alpen-Adria Universität Klagenfurt, Austria
Mittermeir, Roland (Co-chair)	Alpen-Adria Universität Klagenfurt, Austria
Schubert, Sigrid	Universität Siegen, Germany
Sysło, Maciej M.	Nicolaus Copernicus University, Toruń and University of Wrocław, Poland
Verhoeff, Tom	Eindhoven University of Technology, The Netherlands

Additional Reviewers

Alimisis, Dimitris	School of Pedagogical and Technological Education, Patras, Greece
Antonitsch, Peter	Alpen-Adria Universität Klagenfurt, Austria
Bezáková, Daniela	Comenius University, Bratislava, Slovak Republic
Boytchev, Pavel	Sofia University, Bulgaria
Forišek, Michal	Comenius University, Bratislava, Slovak Republic
Fuchs, Karl	Universität Salzburg, Austria
Haberman, Bruria	Holon Institute of Technology and Davidson Institute of Science Education at The Weizmann Institute of Science, Israel
Hubwieser, Peter	Technische Universität München, Germany
Huizing, Kees	Eindhoven University of Technology, The Netherlands
Jevsikova, Tatjana	Vilnius University, Lithuania
Kabátová, Martina	Comenius University, Bratislava, Slovak Republic

Kołczyk, Ewa	University of Wrocław, Poland
Koncilia, Christian	Alpen-Adria Universität Klagenfurt, Austria
Kubincová, Zuzana	Comenius University, Bratislava, Slovak Republic
Moro, Michele	University of Padua, Italy
Motschnig, Renate	Universität Wien, Austria
Proulx, Viera K.	Northeastern University, Boston, MA, USA
Romeike, Ralf	Universität Potsdam, Germany
Salanci Ľubomír	Comenius University, Bratislava, Slovak Republic
Semenov, Aleksej	Moscow Institute of Open Education, Russia
Sendova, Jenny	Bulgarian Academy of Sciences, Sofia, Bulgaria
Sperry, Carol	Millersville University, USA
Tomcsányiová, Monika	Comenius University, Bratislava, Slovak Republic
Vaníček, Jiri	École Polytechnique Fédérale de Lausanne, Switzerland
Weger, Benne de	Eindhoven University of Technology, The Netherlands
Weigend, Michael	Westfälische Wilhelms-Universität Münster, Germany
Winczer, Michal	Comenius University, Bratislava, Slovak Republic
Zur, Ela	The Open University of Israel, Raanana, Israel

Organizing Committee

Ivan Kalaš (chair), Daniela Bezáková, Andrea Hrušecká, Roman Hrušecký, Ľudmila Jašková, Martina Kabátová, Katarína Kalašová, Zuzana Kubincová, Katarína Mikolajová, Renáta Odnechtová, Roman Riška, Monika Tomcsányiová, Mário Varga; all Department of Informatics Education, Comenius University, Bratislava, Slovakia.

Main Sponsors

- Slovak Society for Computer Science, Slovak Republic
- Bundesministerium für Unterricht, Kunst und Kultur, Vienna, Austria
- Datalan a.s., Solvak Republic

Table of Contents

Informatics Education – The Spectrum of Options

National Perspectives

Outreach Programs

Teacher Education

Informatics in Primary Schools

Advanced Concepts of Informatics in Schools

Competitions and Exams

Wild Programming – One Unintended Experiment with Inquiry Based Learning

Pavel Boytchev

KIT, Faculty of Mathematics and Informatics,
Sofia University, blvd J. Bourchier 5, 1164 Sofia, Bulgaria
`boytchev@fmi.uni-sofia.bg`

Abstract. This paper describes one unplanned experiment of a 6th grade student writing her first computer program for 3D graphics before learning any programming language. Some intriguing aspects in her program are analyzed, especially the emerging understanding of key concepts like enumeration, naming conventions of variables and symmetry in 3D space. The paper also identifies two main directions of mental processes. The first direction is actively supported by the school. It is based on presenting and using knowledge in a distilled error-free way. The other direction encompasses techniques needed to identify wrong solutions and to find a way to overcome problems and reach a correct solution. This direction in underrepresented in the educational system and it is left uncultivated as a result. Students are expected to develop such skills by themselves.

Keywords: Programming, cultivated education, emerging understanding.

1   About Wild and Cultivated Strawberries

Many people like strawberries, especially the ones that are big, juicy and tasty. These are the cultivated strawberries. The wild strawberries are completely different – they are small, plain, but extremely fragrant. Wild strawberries are perfect for making strawberry jam. Almost three hundreds years ago the French person Amédée-François Frézier brought the wild Chilean strawberry *Fragaria chiloensis* to Europe. When hybridized with the North American *Fragaria virginiana*, it gave birth to the modern garden strawberry [1].

Nowadays, some people are surprised that wild strawberries can be eaten. They don't expect that a wild fruit can be edible. So far they have only tasted cultivated strawberries, properly wrapped and labeled.

It appears that the cultivation of strawberries has a common ground with the cultivation of … people. For centuries learning and teaching are tightly bound to this cultivation. The situation leads to the question whether we have reached the status of believing that this cultivation is inherent to education.

When we give a toy to a child, we just show quickly how it is used. Then the child continues to play with the toy and to explore its functions. This is a kind of "wild learning". The situation in the classroom is much more cultivated. Everything is being thoroughly premeditated and explained. To some extent this attenuates the natural

I. Kalaš and R.T. Mittermeir (Eds.): ISSEP 2011, LNCS 7013, pp. 1–8, 2011.

pursuit of wild experimenting. Within the cultivated education, students see only the correct way of solving a problem or undertaking a research. They are detached from the wild exploration, where mistakes are the driving force of learning. People learn from their mistakes – mistakes are as educational as non-mistakes [2]. Unfortunately, we want to exclude all mistakes and even chances of mistakes from the learning process.

Let us consider as an example the discipline *Computer Science* and focus on one of its subdisciplines – *Programming*. The education in *Programming*, independent of the programming language being studied, follows a canonical methodology, which leads to a cultivated, but a sterile state. Is it possible for a student to learn something in this way? Yes, it is, this is the "normal" way of learning things and a lot of people learned to use a programming language in exactly this way. The question is whether wild learning is also applicable in this context. What would happen if students are given only a primary explanation and then they are left alone to experiment with the programming? Would it be possible for complex and abstract concepts in *Programming* and *Computer Science* to emerge? If we forget about the canonical mythology and provide educational freedom, would this lead in a natural way to blending elements from different disciplines?

2   The Experiment

The experiment happened in a casual day, while we were engaged with reviewing more than a hundred multimedia projects written by students from 5th to 7th grades. As expected the projects were highly varied. There were PowerPoint presentations, frame-by-frame-hand-drawn video clips accompanied by personal poetry and even a few animations programmed in OpenGL.

A 6th grader saw the projects and became extremely interested. After seeing several multimedia projects, she said curtly: *Why do we not study how to do this at school? Why do we learn only Paint, Excel and Word?* The reply to these rhetorical questions was that school is not the only place where we can learn new things. Then she asked how she could make some cute animation ... *not something recorded by a camera, but animation that is entirely computer-generated.*

There was a big hesitation whether to tell her about Elica – the programming environment used to build many other educational applications including applications within the frame of three European projects – *DALEST* [3,4], *InnoMathEd* [5] *and Fibonacci* [6]. The main problem was that the girl had never done any programming. She had never written a single command in a programming language, so diving directly into the world of programmed 3D animations could be a disaster. On the other hand, it was a unique moment that she explicitly expressed her strong will to learn something that goes far beyond the school curriculum.

Thus the casual lesson started with some quick introduction to 3D coordinates. The girl was not aware of the Cartesian 3D coordinate system, but she had studied the 2D coordinate system at school. When she was asked *Do you recall 2D coordinates* she answered *Yes*, wrinkling her forehead. It was like just this single question that made her step back regretfully. However, we used the two edges of the desk as X and Y axes, and an upright pen as Z axis in order to model a coordinate system. After a moment, while placing hands on desk surface, the girl proudly said that X and Y were forming *a flat plane*.

It was time to move to the next step – introducing coordinates. The girl was shown the approximate positions of objects with coordinates (10,0,0) and (0,0,10); and then she was able to point in the space the positions of (0,10,0), (10,10,0) and (10,10,10). She was even asked to point (10,-10,-10) and after few seconds of hesitation she placed her hand below the desk in the correct position in respect to the axes. It was surprising how fast she managed to get oriented in the 3D space, so it was time to make the final step – writing a true computer program.

For this step we used Elica. Its acronym stands for *Educational Logo Environment for Creative Activities*. Although it is based on Logo, a language largely and wrongly assumed to be childish, Elica provides support for object-oriented, functional and procedural programming – all at the same time. It was quite risky to ask a child that had absolutely no programming experience to write a program. Thus, hoping to make just a "presentation", we showed her a simple program that draws and rotates two cubes. A snapshot of the screen, together with the program code is shown in Fig. 1. The make statements define the cubes and their properties, and demo is "responsible" for the rotation.

The most surprising element in this program was when the girl was asked to give names to the cubes. She was curious why, but she accepted without problems that all objects in the animation must have their own unique names. In this way she could "touch" the objects and "tell" them what to do. Most likely the problem with naming was that in Paint the picture is not composed of individual entities, but is treated as a single piece of painted nameless strokes.

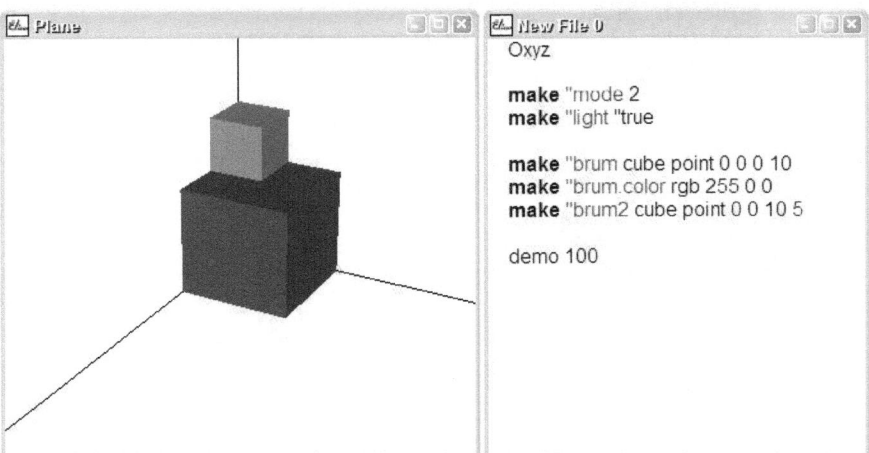

Fig. 1. The program for creating and rotating two cubes

Anyway, the girl decided that the cubes must be called brum and brum2 (echoic words corresponding to *whirr* or *buzz*). We did not influence this decision and we did not discuss it with her.

The experiment up to this point was about 5-10 minutes long. The final explanation that we provided was that Elica could use not only cubes, but spheres, cones, and many other shapes. After this note the girl was left along.

3   The Result

Approximately 15 minutes later we went to her room to see what is going on and we were shocked to see a panda on the computer screen as displayed by the snapshot in Fig. 2. This panda was the first program ever of this 6th grader! It was so unbelievingly well done, that we immediately studied it and asked several question:

We: *How do you know how to use spheres?*

Girl: *You told me that I can use spheres, so I looked for "сфера" (i.e.* sphere *in Bulgarian) in Google and found that in English it is "sphere". So I just used this word and everything worked so well.*

We: *Did you try other objects?*

Girl: *Yes, but they didn't work out.*

We: *Yes, to construct them you need more numbers, because these objects are more complex.*

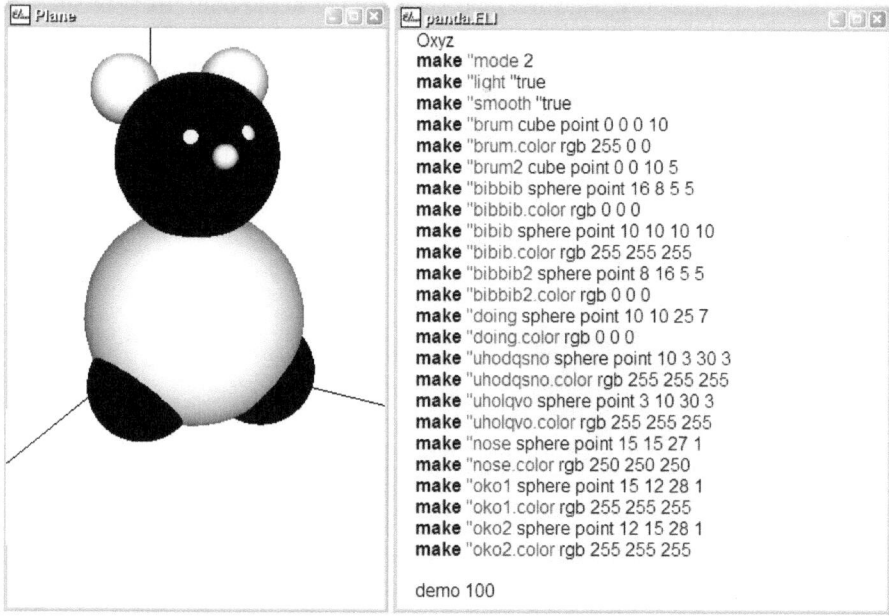

```
Oxyz
make "mode 2
make "light "true
make "smooth "true
make "brum cube point 0 0 0 10
make "brum.color rgb 255 0 0
make "brum2 cube point 0 0 10 5
make "bibbib sphere point 16 8 5 5
make "bibbib.color rgb 0 0 0
make "bibib sphere point 10 10 10 10
make "bibib.color rgb 255 255 255
make "bibbib2 sphere point 8 16 5 5
make "bibbib2.color rgb 0 0 0
make "doing sphere point 10 10 25 7
make "doing.color rgb 0 0 0
make "uhodqsno sphere point 10 3 30 3
make "uhodqsno.color rgb 255 255 255
make "uholqvo sphere point 3 10 30 3
make "uholqvo.color rgb 255 255 255
make "nose sphere point 15 15 27 1
make "nose.color rgb 250 250 250
make "oko1 sphere point 15 12 28 1
make "oko1.color rgb 255 255 255
make "oko2 sphere point 12 15 28 1
make "oko2.color rgb 255 255 255

demo 100
```

Fig. 2. A 3D panda – the girl's first program. The long sequence of make statements suggests the application of some complex programming concepts.

There were some surprising things in the program. The first objects that the girl added to the cubes had funny meaningless names, like bibbib and doing (again echoic words). Then she started to embed sense in the names, the panda ears were named uhodqsno (right ear) and uholqvo (left ear), the nose was called nose (in English!).

And then suddenly she jumped to a numerical notation, which generates shorter names and is the doorstep to enumeration – oko1 (eye 1) and oko2 (eye 2). Enumeration is a key programming concept, which is the core of arrays, cycles and iterations. It is unexpected to observe similar transitions at such an early stage.

Another interesting observation, realized several days later, was the use of symmetry. If we were to make a panda, we would orient it along some of the axis, so that the whole panda body is symmetrical in respect to a trivial vertical plane (like the plane y=0). This would make it much easier to position symmetrical body parts like eye, ears and legs. If one part has coordinates (x,y,z), then its symmetrical part would be at (x,-y,z).

However, the girl's panda was not oriented in a way to use such an idea, yet it was completely based on symmetry – the symmetry plane was the bisecting plane x=y. This plane makes points (x,y,z) and (y,x,z) symmetrical.

Some of the symmetrical coordinates are shown in Fig. 3. The spheres for the ears (the statements that create variables uhodqsno and uholqvo) are placed at (10,3,30) and (3,10,30). The centers of the eyes (oko1 and oko2) are at (15,12,28) and (12,15,28).

The 3D objects that the girl created were appended to the definitions of the two cubes. When the panda bear turned the cubes were poking out of her lower back – see Fig. 4. It looked like these leftovers were the first ever programming bug of the girl, but this conclusion was premature and … wrong. The girl explained us that *these cubes are the chair of the panda and that everything is correct!!!*

```
make "doing.color rgb 0 0 0
make "uhodqsno sphere point 10 3 30 3
make "uhodqsno.color rgb 255 255 255
make "uholqvo sphere point 3 10 30 3
make "uholqvo.color rgb 255 255 255
make "nose sphere point 15 15 27 1
make "nose.color rgb 250 250 250
make "oko1 sphere point 15 12 28 1
make "oko1.color rgb 255 255 255
make "oko2 sphere point 12 15 28 1
make "oko2.color rgb 255 255 255
```

Fig. 3. Close-up of some symmetrical coordinates

Later on the same day the girl made another program – a face of a child with lips, eyes with irises, nose and hair. We showed her some simple form of animation like inflating and deflating the face by changing one of its radii. It was quite interesting how the girl "accepted" that a sphere had actually three radii – one along each of the axis; and by making them non-equal we could deform the sphere – and the girl quickly completed the sentence for us – *into an egg.*

4 Afterthoughts

The result of this experiment showed that programming is not hard at all if we do not insist to tell all details and provide complete scientifically correct explanations. A child can start programming without understanding *everything about the program*. This method is very close to the exploration of an unknown toy, when the child is left to experimentally find out what can be done.

Additionally, letting a student play with and in (!) a programming environment does not impose any restrictions to imagination. While creating something entirely by her, the 6th grader freely integrated art activities with programming. If an adult was about to write his/her first program for 3D graphics, he/she would most likely start with something more conventional, more systematic ... or even more cultivated (like reading the documentation).

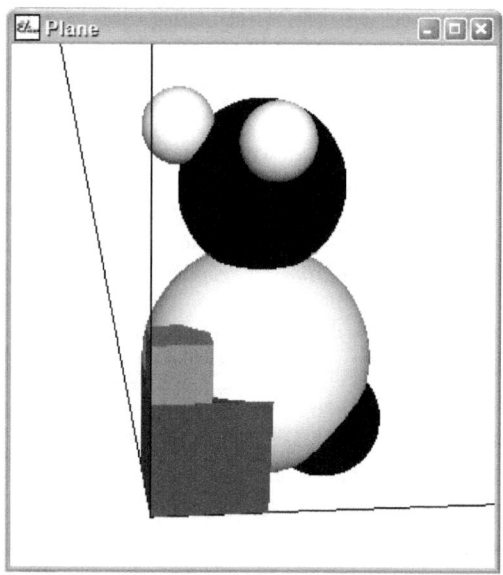

Fig. 4. There is no bug here, but the chair of the panda

The experiment shows one of the advantages of the programmable educational environments. In such environments students have at their disposal instruments for describing not only what they do, but also the individual steps of their constructions. Students' programs, independent on their complexity or simplicity, are projection of students' thoughts. Even "the most innocent" elements like the selected naming convention of variables, provide clues about the existence of specific skills and the level of understanding of key concepts.

Cognitive psychology explores various types of thinking. Two of the most distinguished types are the *vertical thinking* and the *horizontal (lateral) thinking* [7].

Some of the main features of both thinking types as identified by Paton [8] based on [9] are listed in Table 1. The cultivated approach in education fits perfectly to the vertical thinking, while the wild approach – to the horizontal one.

Table 1. Vertical and horizontal thinking mapped to cultivated and wild education

Feature	Vertical thinking	Horizontal (lateral) thinking
Characteristics	selective, analytical	generative, provocative
Focus on	correctness	richness
Individual steps	must be always correct	some could be wrong
Negative experience	blocks off certain pathways	does not exist
Thinking process	finite	probabilistic

Doing research by writing a computer program reveals much more information if we focus not only on the final program as a static artifact, but also on the program's evolution from scratch till the end, passing through many incomplete and buggy states. This evolution shows a new class of thinking and is indicative for the path of gaining concrete skills and understanding key threshold concepts. The *horizontal thinking* is the one which happens when students stumble upon a wrong solution and try to transform the solution to a correct solution. This thinking helps the students to "feel" when a research is going in the wrong direction before it is too late. This is the thinking that allows the students to attempt different solving strategies over a problem instead of being blocked off by failures.

Educational environments that allow experimentation via programming develop not only the vertical, but also the horizontal thinking. A programming description of a solution is rarely written perfectly from the very beginning. Often it is required to remove bugs or to improve some existing elements. Debugging and optimization are some of the processes that develop horizontal thinking. Unfortunately, horizontal thinking is not taught at school, but is expected to be acquired by the person on her or his own This shows one visible discrepancy between what is taught and what is expected to be learned. The vertical thinking is completely cultivated up to the level of lack of critical thinking – here is a problem, here is an algorithm for solving it, follow the algorithm and you will get a correct solution. At the same time the horizontal thinking is growing in the wild, uncontrolled and undirected.

Would it be better to restore the balance between both thinking types? Could we make the vertical thinking wilder (i.e. to make it more independent and more creative by deframing students' thinking and letting them experiment)? Or could we make the horizontal thinking at least more cultivated (i.e. to help students to analyze wrong situations and developing skills for searching new solutions)? These are questions that need yet to be answered.

5   As an Epilogue

The experiment described in this paper was not planned, that is why it was not possible to observe the process of the creation of the panda. Only one student was involved, so it is too early to draw general conclusions. It is not known whether the wild programming always leads to small aromatic fruits or the result was pure fortuitous event. Maybe wild programming is not applicable to mass education? Maybe it is more suitable for individual learning? The answers of these questions are unknown, but the thing, which is known is that without the efforts of Amédée-François Frézier, today, three hundreds years later, it would be impossible to enjoy the garden strawberry. And something else is also known. Frézier not only brought the strawberry to Europe, but he was the mathematician whose works laid the fundaments of the 3D geometry in military construction and engineering.

As for the usage of digital technologies in education, the Logo-philosophy (a main topic in the international conference Constructionism 2010 [10]) is not to focus only on the informational or the technological sides, but to fully explore the potential of students to be constructors of their knowledge, to learn through inquiry and to share their works.

References

1. Darrow, G.: The Strawberry: History, Breeding and Physiology, Chapter 4: The Strawberry From Chile. Holt, Rinehart and Winston, New York (1966)
2. Boytchev, P.: Pedagogical Inversion. Presented at the 4th International Conference for Theory and Practice in Education, Budapest (2011)
3. Boytchev, P., Chehlarova, T., Sendova, E.: Enhancing Spatial Imagination of Young Students by Activities in 3D ELICA Applications. In: 36th Spring Conference of the Union of Bulgarian Mathematicians, Varna, pp. 109–119 (2007)
4. DALEST project,
   http://www.elica.net/site/museum/Dalest/dalest.html
5. InnoMathEd project,
   http://www.math.uni-augsburg.de/prof/dida/innomath
6. Fibonacci project, http://fibonacci.uni-bayreuth.de/
7. Robertson, S.: Types of thinking. Routledge, London (1999)
8. Paton, B.: Lateral Thinking,
   http://www.solutioneers.net/solutioneering/lateralthinking.html
9. Bono, E.: Lateral Thinking: A textbook of creativity. Penguin, Harmondsworth (1977)
10. Clayson, J., Kalas, I. (eds.): Proc. of Constructionism 2010, Paris, France. Comenius University, Bratislava (2010)

Informatics Education for New Millennium Learners

Valentina Dagienė

Vilnius University, Faculty of Mathematics and Informatics,
Naugarduko str. 24, Vilnius, LT-03223, Lithuania
`valentina.dagiene@mif.vu.lt`

Abstract. The stage of the 21st century education technologies is specified by the technology development level, dependence of economics on information technologies and computer network progress. Technical and educational training aids have a pronounced influence on the training activity. Computer instructional aids, computer networks, virtual computer environments perform all the main didactical functions; impart the new worksheet, help to consolidate it, aid in solving problems, accumulates feedback, and locate the learning difficulty. Currently educators are active in defining what 21st century competencies are and how they can be efficiently integrated into existing educational systems. Informatics education was begun at schools of many countries almost 50 years ago, but with the onset of this millennium it slackened due to the need for use of technologies in practice. Now the revival is in process: education policy makers of different countries note that informatics education can markedly improve the competencies of the 21st century learners. The paper deals with these issues and tries to search for possible answers.

Keywords: 21st century competencies, informatics in school, ways to learn informatics, information technology.

1 Introduction

We are living in a global connected world. Through the process of globalisation, our societies have experienced a profound transformation from reliance on an industrial to a knowledge base. More and more are we discussing about 21st century competences: flexibility and the capacity to make creative connections, deep understanding, good team-working, etc. We need to learn to generate, process, and sort complex information, to think critically and systematically, to make decisions according to various circumstances and different forms of evidence, to be adaptive and flexible to a new kind of information, to recognize new phenomena and to deal with them in the best way, to identify and solve real world problems, to be more and more creative.

The Organisation's for Economic Cooperation and Development (OECD) has defined most general 21st century competencies divided into four categories: 1) functioning in socially heterogeneous groups; 2) acting autonomously; 3) using tools interactively; 4) thinking (a "cross-cutting" competency).

Informatics (or computer science), the science of algorithmic processing, representation, storage and transmission of information, is an important discipline in

I. Kalaš and R.T. Mittermeir (Eds.): ISSEP 2011, LNCS 7013, pp. 9–20, 2011.

the knowledge society and should be introduced into secondary or even primary school. Informatics is a fascinating research with a great impact on the real world, full of spectacular ideas and great challenges [1].

2   Short Glance at Teaching and Learning Informatics at School

Informatics as a school subject (in fact, elements of informatics) was introduced in schools in the early 1980s in many countries around the world. The development of informatics in schools, in particular, is characterized by permanent changes in hardware and software and it has some variations from country to country. The didactic approaches and key topics of informatics played a very important role as well.

Informatics, as a separate subject in high or secondary schools, was taught in the majority of East European countries, where fundamental and academic trends of teaching are more prevalent until nowadays. As a compulsory or partly compulsory subject, informatics has been delivered in Belarus, Bulgaria, Czech Republic, Lithuania, Poland, Romania, Russia, Slovak Republic, Hungary, Germany, and other countries [2, 3, 4]. The course is being changed permanently: in the beginning, teaching about computers and training of the programming skills used to get more attention, later a shift to the development of skills of practical use of information and communication technology in teaching and learning were observed, including the focus on technology-enhanced learning in the last decade (Fig. 1). In today's world, more and more educators have started talking about informatics in schools as a language of technology [5], informatics as a bridge between using digital devices and understanding the digital world, informatics as an everyday language for digital native youngsters. It means that informatics education should be deeply rethought.

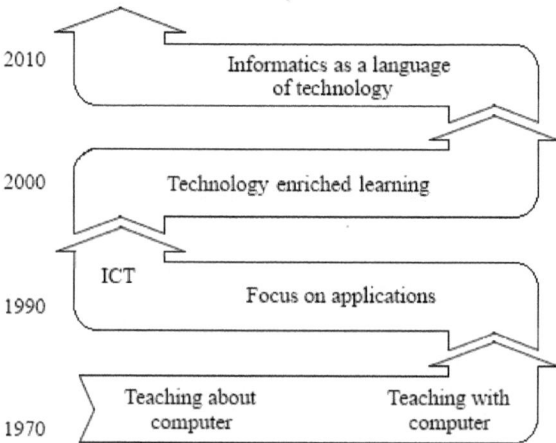

Fig. 1. Changes in teaching informatics as a schools subject during decades – a spiral evolution

All the countries pay a rising attention to ICT implementation in education [6, 7, 8]. Those countries which have informatics as a separate subject usually treat ICT as part of it; however, most of the time in the teaching process ICT is assigned to the technology itself, but not to its application to the process of learning. In order to emphasize the novelty of the course in informatics and the aspect of its applicability, several countries, including Lithuania, have renamed it into information technology. Nonetheless, usually information technology curricula keep some basic elements of informatics education, e.g. understanding of algorithm or combinatorial manipulations.

Concerning the contents of informatics or information technology in primary or secondary school there is almost no common international agreement or accepted IT-framework. Several researchers have been trying to initiate discussions on the issue: What concepts of informatics and information technologies should be included in general education? [1, 9, 10, 11, 12]

It is almost a common provision that the fundamentals of algorithms and programming are the key concepts in school informatics education. Then, what concepts should we include in informatics education besides algorithms and programming? What is the ratio of programming and information technology concepts and their application? How could we use information technology for collaborative learning to represent these concepts for students and ensure productive and sustainable learning? The most important point is how the skills and capabilities that we would like to bring to our students through informatics education agree with the needs of newly rising 21st century learner's.

3   Opportunities for the New Millennium Learner

In 2007, the OECD Centre for Educational Research and Innovation (CERI) launched the New Millennium Learners (NML) project (http://www.nml-conference.be/). It has the global aim of investigating the effects of digital technologies on school-age learners and providing recommendations on the most appropriate institutional and policy responses from the education sector.

The concept of New Millennium Learners suggests that the technology uptake, particularly by younger generations, has an effect on the way people build their identities, communicate socially, and manage information and knowledge. However, the fact that young people are increasingly attached to and knowledgeable in terms of technology does not necessarily mean that they develop by themselves the range of skills and competencies that the knowledge economy requires. Today's children are thought to be flexible with computers, immediate to communicate, creative with technology, and highly skilled at multitasking in a world where ubiquitous connections are taken for granted [13]. Multitasking is the very normal approach to using digital media for everyday lives of our students. Undoubtedly, students' recurrent activity with these technologies shaped their communication, knowledge management, learning, and even their personal and social values.

The so-called New Millennium capabilities or competencies cover the range of skills and competencies that young people will be required to have in order to be efficient workers and responsible citizens in the knowledge society of the 21st century. Many children gained technology manipulation skills intuitively, actually by trying and doing. Our schools should help these students to develop wider and deeper capabilities in different areas and improve their learning skills.

The most important capabilities of the new millennium learners are the skills, knowledge and expertise, which students should master to succeed in work and life in the 21st century. There are many variations on how to define the new millennium learner's skills. A group of business leaders, educators and lawmakers in Massachusetts have elaborated five areas for the 21st century learner's education [14]:

Core academic subjects: reading, world languages, arts, math, economics, science, geography, history, government, and civics.

Interdisciplinary themes to be woven into each subject include global awareness, economic, business and entrepreneur literacy, civic literacy, and health literacy.

Learning and innovation skills include creativity, innovation, critical thinking, problem solving, communication, and collaboration.

Information, media and technology skills required by today's students include information and media literacy, communications and technology literacy.

Life and Career Skills, needed to navigate in today's world, include flexibility, adaptability, initiative and self-direction, social and cross-cultural skills, productivity, accountability, leadership, and responsibility.

In order to summarize the debates on the key competencies, the world institutions, such as OECD, UNESCO, and European Commission launched surveys and settled several recommendations [7, 15, 16, 17, 18].

Discussions have been promoted around the idea that 21st century students will learn to think both critically and creatively, be skilled at working collaboratively, and understand how to take risks constructively. They will learn and understand their connection to the world around them, use technology to pursue research and communicate with others, feel comfortable working in teams and will develop the strength and skills to assume leadership responsibilities.

Digital literacy, media, ICT and other modern technology-based skills are essential requirements for the 21st century learner's education. ICT competency and skills are important for every citizen in a modern society.

The fact that these skills have never been the focus of traditional education is a serious problem. Delivery and acquisition of these skills in teaching and learning to students of primary and secondary education will require a shift in what we teach, how we teach it, the tools we use and how we educate, train, nurture and retain our teachers and school leaders. The overarching challenge for all educators today is to rethink not only what they teach, but "how they empower students to use that information" [19].

A holistic view of 21st century to teaching and learning combines a discrete focus on learner's outcomes (a blending of specific skills, content knowledge, expertise and literacy) with innovative support systems to help students master the multi-dimensional abilities. It is necessary to prepare students for a rapidly evolving global and technological world, and promote innovation through critical thinking, problem solving, collaboration, and technology integration, while building the mastery of the core content and background knowledge.

The OECD/CERI project on new millennium learners and, in particular, the international conference on the 21st century competencies taking place in Brussels in September 2009 (http://www.nml-conference.be) carried out a survey in many countries [20]. A three dimensional model was elaborated to conceptualise the competencies for 21st century learners: 1) information, 2) communication, 3) ethics and social impact (Fig. 2).

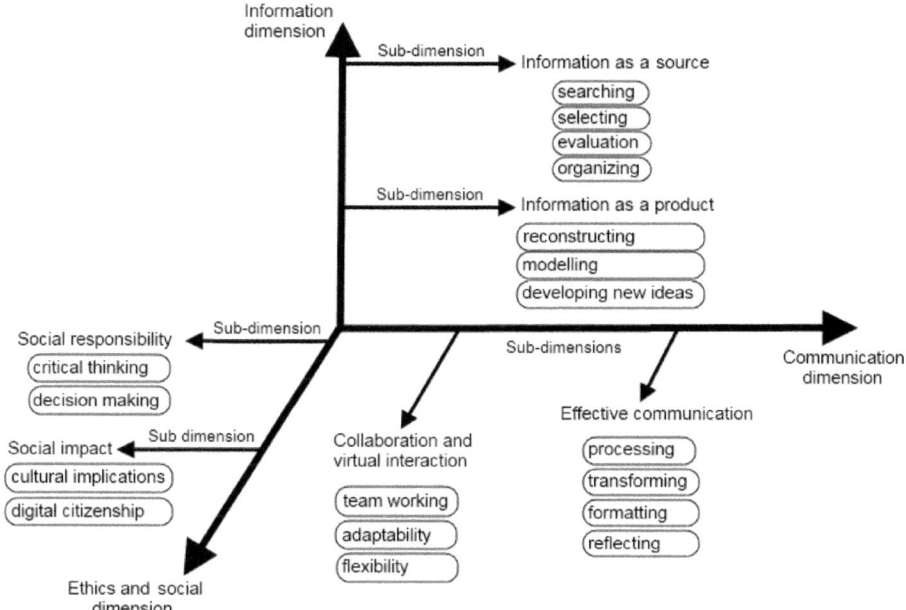

Fig. 2. The three-dimensional model of the new millennium learners' competencies

The information dimension covers research and problem solving skills as they both involve, at some point, defining, searching for, evaluating, selecting, organising, analysing, and interpreting information. This dimension includes two sub-dimensions: firstly information as a source (searching, selecting, evaluating and organising information), and secondly information as a product (restructuring and modelling of information, the development of one's ideas or knowledge).

The communication dimension involves ICT applications and reinforces the development of skills of coordination and collaboration between peers. This dimension also has two sub-dimensions: 1) effective communication and 2) collaboration and virtual interaction.

The third dimension, related to ethics and social impact, is connected with globalisation, multiculturalism, digital citizens of the 21st century. As in the previous dimensions, there are two ethical sub-dimensions: social responsibility, and social impact.

Generally summarising the 21st century competencies could be characterized as the skills, knowledge, attitudes and values that all people need to know and be able to do in order to live meaningfully in, and contribute to a well functioning society. Various countries stressed slightly different competencies. However, there are many competencies mentioned by almost all countries in their general education (formal or informal). Many of these modern competencies are based on technology enriched learning and have (or should have) deep relations with informatics (Table 1).

Table 1. New Millennium Learner's competencies and their relation to informatics education

Competence	Related concepts that can be delivered through informatics education
Creativity/ innovation	Hardware and software design, information visualization, development of algorithms, programming
Critical thinking	The limits of computations (what computers cannot do), data analyses
Problem solving	Modelling, structures, notion, problem decomposition, algorithmisation, automation
Decision making	Artificial intelligence, parallelization
Communication	Computer networks, mobile technology
Collaboration	Networking, sharing approaches
Information literacy	Handling information, objects, data bases and information retrieval
Research and inquiry	Abstraction, formalisation, modelling
Media literacy	Graphics, simulation
Digital citizenship	Social, ethical, legal issues, internet security, privacy, data processing
ICT operations	Computer logics, dealing with information
Flexibility and adaptability	Patterns and recognitions
Initiative and self-direction	New devices and new computer programs
Productivity	Computer applications, networks, information organisation, automation
Leadership and responsibility	Roles in virtuality, robotics

Most countries integrate the development of 21st century skills and competencies in a cross-curricular way, i.e., across the subject areas. ICT-related skills are often the exception to this, i.e. they are taught in some countries as a separate subject.

4   Some Modern Ways of Learning Informatics at School

Twenty-first century standards, assessments, curricula, instruction, professional development and learning environments must be aligned to produce a support system that produces the 21st century outcomes for today's learners. Although informatics is recognized by professionals as a field with scientific and engineering orientation, in the high school education system informatics is considered as a non-important subject [21].

Understandably, there is an intense competition between subjects (as well as topics within subjects) in school: traditional subjects, which have been taught for centuries, are in a better position than the new rising ones. Nobody argues the importance of chemistry or physics in school; however, the science of information or knowledge technology still has not been recognised as a school subject. Concepts of informatics play a central role in all curricula and standards for informatics education at school. However, in practice at school the training of skills in software application is very often given much more room than the understanding of fundamental concepts of informatics.

The "concept" can be understood as extensive information on a particular object, existing in the human sense. In formal sciences "concept" is defined as an abstract idea which generalizes separate objects, defines attributes and relations between objects. The content of a concept can vary a lot as it depends on personal experience. Concepts of informatics are tightly related with our intensions (what we would like to teach at school) and expectations of a modern knowledge society.

Researchers identified many ideas about what informatics is and what it could be in K-12 classrooms [1, 9, 11, 22]. In his research Hromkovic presents components that represent the basics of informatics and should be taught at school [1]. They include programming, computability and automata theory. The automata theory (as well as graph theory) can be visually represented by simple schemes; it can be used to present many examples from everyday life. The automata theory, for example, can be considered as part of the structure and pattern concept presented above.

Starting with using computer applications (e. g., educational aids, learning objects), learners could become not just tool users but tool builders (Fig. 3). Dealing with information in the knowledge society, we need a set of concepts, such as abstraction, recursion, and iteration, to process and analyze data, and to create real and virtual artefacts. Informatics should be seen as a problem solving methodology that can be automated, transferred and applied across subjects.

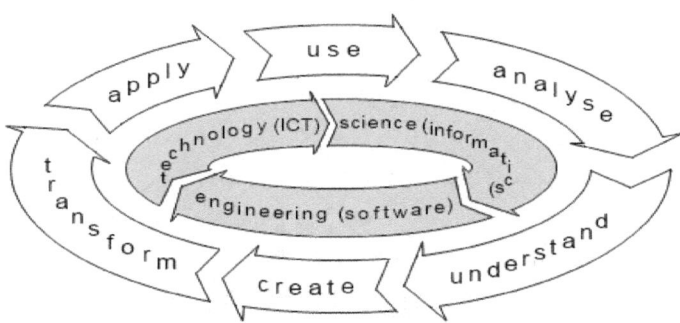

Fig. 3. The today's learning growth – "from technology through science to engineering"

Curricula and standards for secondary school describe the learning contents and methods of learning. Some international guidelines have been developed in the field of informatics that define for a larger group of countries which content areas and which way of learning can be appropriate, e.g., the UNESCO/IFIP curriculum [23], the ACM K-12 curriculum [24].

The ACM K-12 curriculum report references the idea of information technology fluency of the National Research Council and describes the informatics-concepts as ten basic ideas that underlie modern computers, networks and information. A computer-fluent student would master information technology on three orthogonal axes: concepts, capabilities, and skills. Concepts are understood as the ten basic ideas that underlie modern computers, networks, and information: 1) computer organization, 2) information systems, 3) networks, 4) digital representation of information, 5) information organization, 6) modelling and abstraction, 7) algorithmic thinking and programming, 8) universality, 9) limitations of information technology, and 10) societal impact of information technology.

The ACM Model Curriculum for K-12 [24] provides a definition of informatics specifically for K-12 educators. Computer science (informatics), it argues, is neither programming nor computer literacy. Rather, "it is the study of computers and algorithms processes including their principles, their hardware and software design, their applications, and their impact on society" (p. 1). We can start talking about informatics concepts: 1) computer organization (hardware and software design, computer security), 2) information systems (applications in information technology and information systems), 3) networks (collaboration), 4) digital representation and visualisation of information (graphics), 5) information organization (data bases and information retrieval), 6) modelling and simulation (virtuality, artificial intelligence), 7) algorithmic thinking and programming (languages and paradigms), 8) universality (translation between levels of abstraction), 9) limitations of information technology, and 10) societal impact of information technology (internet security, privacy, intellectual property, ethics, impact on society, etc.). These informatics concepts are very significant in a knowledge society and have tight relations with the 21st century learners' competencies (Fig. 4).

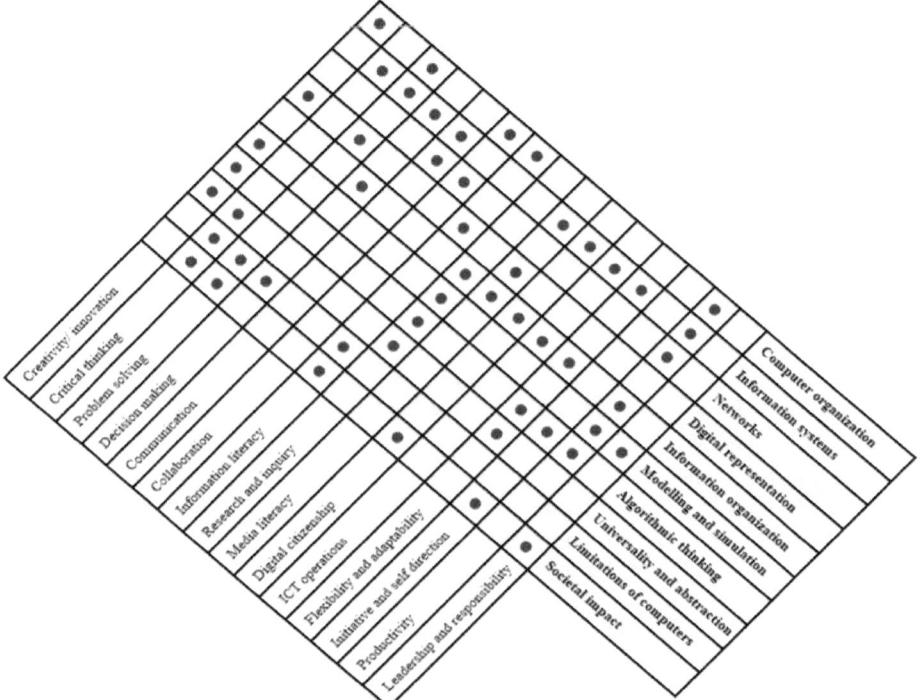

Fig. 4. Fundamental informatics concepts and new millennium learners' competencies

The standard developed by the German Society for informatics GI [25] is quite new and has fresh ideas for informatics education in secondary schools, grades 5 to 10. The GI standard proposed two main areas for teaching informatics: content area and process area. Each content area can be combined with each process area together with examples of typical tasks that are suitable for secondary school education. The content part covers five basic concepts: a) information and data, b) algorithms, c) languages and automata, d) informatics systems, e) informatics, man and society, while the process area promotes actions combining with concepts, e.g. modelling and implementation, representation and interpretation, structuring and networking, communicating and cooperating, arguing and evaluating. More detailed initiatives in terms of reviewing and structuring informatics education are presented in the paper by Peter Micheuz [11].

The power of informatics is that it applies to every other type of reasoning. While learning the informatics students would understand that problems can be solved in multiple ways, have reasonable expectations about the prospect of producing a working solution.

There are many ways how to bring informatics education to high school. In order to harmonise with the educational system, we should agree on fundamentals of an informatics curriculum (or framework) and establish a balance between teaching and learning (Fig. 5).

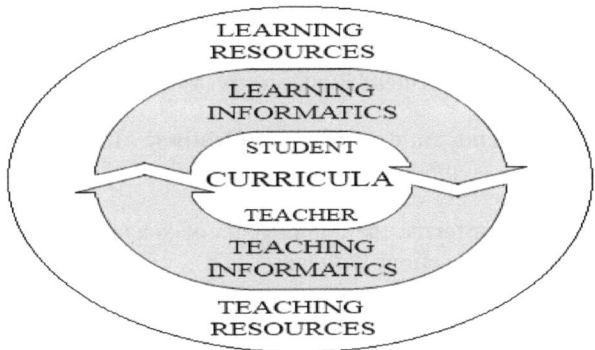

Fig. 5.

Kalas and Winczer suggested two main ways to support informatics education: provide modern university pre-service teacher development and produce attractive and inspiring informatics textbooks, educational software and learning resources for children, students and teachers [26].

We cannot change how our students learn, unless our teachers are equipped to teach in new ways. Studies show that a teacher's qualification has a significant effect on student's performance, more than any other variable (e. g. [27]). It is unreasonable to expect that our students will ever gain the skills and knowledge to succeed in the 21st century, if they are taught primarily by the educators trained using a model developed in the 19th century. It is necessary to rethink and overhaul the teacher training and professional development programs, in order to recruit and retain high achieving educators who have up-to-date knowledge of 21st century skills.

To improve teachers' (e. g., of informatics) competencies, the methods of action research – a powerful tool for a change and improvement at the local level – has been suggested [28]. The main goal of the method was to help students engage in and discover fundamental issues, concepts and problems of informatics through exploratory learning and open informal collaborative discussions [18].

Bringing informatics in a formal track to schools by means of curricula is quite important. However, it is necessary to support the informal ways of introducing students to informatics. So another way to bring informatics to school can be through developing attractive activities based on informatics concepts. Contests are among them. Contests are exceptionally valuable for motivating and involving pupils in computer science [29].

The International Contest on Informatics and Computer Fluency (named Bebras in Lithuanian, or Beaver in English, www.bebras.org) can be an example of bringing informatics concepts to students in an informal way. The Bebras contest started in a coordinated way: running contests at schools, where solutions may be submitted to some central authorities or some local organizers.

Any contest needs a challenging set of tasks. The Bebras tasks' developers are seeking to choose interesting tasks (problems) for motivating students to deal with computer science and to think deeper about technology. Collaboration in developing Bebras tasks during international workshops reveals six concepts significant for general informatics education [12, 30]:

- **Information:** the conception of information, its representation (symbolic, numerical, graphical), encoding, encrypting;
- **Algorithms:** action formalization, action description according to certain rules;
- **Computer systems and their application:** interaction of computer components, development, common principles of program functionality, search engines;
- **Structures and patterns:** the components of discrete mathematics, elements of combinatorics and actions with them;
- **Social effect of technologies:** cognitive, legal, ethical, cultural, integral aspects of information and communication technologies;
- **Informatics and information technology puzzles:** logical games, mind maps, used to develop technology-based skills.

It has been agreed that on some of the main concepts to be taught in general education, e.g. algorithms and programming (as a separate or integral part of algorithm construction) is one of the most important concepts of informatics. It could be decomposed into important smaller concepts, e.g. data, variable, cycle, procedure, object, class, etc. Structures and patterns are also important concepts in informatics. The concept "information" undoubtedly belongs to the scope of informatics and information technology.

Computer systems are more difficult to describe (even the concept title itself "computer systems" is not unambiguous, it can be understood as an application of information systems, but not on theoretical grounds). When the concept is not clear enough, it becomes difficult to use and especially to teach.

The social aspect of technology is not an unambiguous concept, which cannot clearly be considered as a separate concept of informatics. No doubt that this topic is very important in our society, but there are still not enough educational examples and systematization for this topic in practice and research.

An important issue is how we present the main informatics and information technology concepts to students. Puzzles and logical games could help attract students, raise their motivation. So, they should be used to express core scientific concepts.

5 Conclusion

Nowadays many countries are seeking to establish the system and educational content for informatics education. It is in progress. Identifying and seeking a solution for the problems related to informatics education and attributed to its implementation under various systems and too much emphasis on technology application education were prompted by the questions. Can information technology education help to solve problems in learning and daily life without education on the principles of informatics education? What specific contents does informatics contain within the different grades? On which fields (e.g., processing information, algorithms, programming, databases, systems design) is mostly placed value? What about the ratio between practice and theory? What could be done to fill the gap between school graduation and the beginning of study (according to education level)? What informatics concepts we need to bring to 21st century learners to make them comfortable and competitive in the knowledge society?

The relatively short history of informatics subject makes classifying it into an independent subject very difficult. Therefore, more research on informatics education should be carried out to justify the need of learning informatics fundamentals and searching new ways and approaches to promote informatics education in classrooms.

References

1. Hromkovic, J.: Contributing to General Education by Teaching Informatics. In: Mittermeir, R.T. (ed.) ISSEP 2006. LNCS, vol. 4226, pp. 25–37. Springer, Heidelberg (2006)
2. Sendova, E., Azalov, P., Muirhead, J. (eds.): Informatics in the Secondary School – Today and Tomorrow. Sofia (1995)
3. Hawkridge, D.G.: Educational Technology in Developing Nations. In: Plomp, T., Ely, A.D. (eds.) International Encyclopaedia of Educational Technology, 2nd edn., pp. 107–111. Pergamon, Great Britain (1996)
4. Dagienė, V.: The Road of Informatics. TEV, Vilnius (2006)
5. Cohen, A., Haberman, B.: Computer Science: A Language of Technology. ACM SIGCSE Bulletin 39(4), 65–69 (2007)
6. OECD. Schooling for Tomorrow. Learning to Change: ICT in Schools. Education and Skills. OECD publ. Paris, OECD Center for Educational Research and Innovation (2001)
7. OECD. Are the New Millennium Learners Making the Grade? Technology Use and Educational Performance in PISA. OECD publications. Paris, OECD Center for Educational Research and Innovation. (2010)
8. OECD. PISA 2009 Results: Executive Summary (2010)
9. Schubert, S., Taylor, H. (eds.): Secondary Informatics Education. Special Issue of Education and Information Technologies, vol. 9(2). Kluwer Acad. Pub., Boston (2004)
10. Micheuz, P.: 20 Years of Computers and Informatics in Austria's Secondary Academic Schools. In: Mittermeir, R.T. (ed.) ISSEP 2005. LNCS, vol. 3422, pp. 20–31. Springer, Heidelberg (2005)

11. Micheuz, P.: Harmonization of Informatics Education – Science Fiction or Prospective Reality? In: Mittermeir, R.T., Sysło, M.M. (eds.) ISSEP 2008. LNCS, vol. 5090, pp. 317–326. Springer, Heidelberg (2008)
12. Dagiene, V., Futschek, G.: Introducing Informatics Concepts through a Contest. In: IFIP Working Conference: New Developments in ICT and Education. Universite de Picardie Jules Verne, Amiens (2010), `http://publik.tuwien.ac.at/files/PubDat_186636.pdf`
13. Pedro, F.: The New Millennium Learners: Challenging our Views on ICT and Learning. OECD-CERI (2006)
14. School Reform in the New Millennium: Preparing All Children for 21st Century Success. Recommendations from the Massachusetts Board of Elementary and Secondary Education's Task Force on 21st Century Skills. Massachusetts Department of Elementary and Secondary Education's (2008)
15. Ananiadou, K., Claro, M.: 21st century skills and competences for New Millennium Learners in OECD countries. In: OECD Education Working Papers, vol. (41). OECD Publishing (2009), doi:10.1787/218525261154
16. Commission of the European Communities Improving competences for the 21st Century: An Agenda for European Cooperation on Schools. Brussels, SEC(2008) 2177 (2008), `http://ec.europa.eu/education/school21/sec2177_en.pdf`
17. Key Competencies. A developing concept in general compulsory education. Eurydice. Survey 5, Brussels (2007), `http://www.eurydice.org`
18. UNESCO. ICT Competency Standards for Teachers: Competency Standard Modules (2008), `http://unesdoc.unesco.org/images/0015/001562/156207e.pdf`
19. Murnane, R., Levy, F.: The New Division of Labor: How Computers Are Changing the Way We Work. Princeton University Press and Russell Sage Foundation (2004)
20. OECD. NML Country Survey. 21st Century Skills and Competencies for New Millennium Learners in OECD Countries (2009)
21. Haberman, B.: Teaching Computing in Secondary Schools in a Dynamic World: Challenges and Directions. In: Mittermeir, R.T. (ed.) ISSEP 2006. LNCS, vol. 4226, pp. 94–103. Springer, Heidelberg (2006)
22. Hromkovic, J.: Algorithmic Adventures: From Knowledge to Magic. Springer, Heidelberg (2009)
23. Anderson, J., Weert, T.: Information and Communication Technology in Education. A Curriculum for Schools and Programme of Teacher Development. Division of Higher Education, UNESCO (2002)
24. Tucker, A., McCowan, D., Deek, F., Stephenson, C., Jones, J., Verno, A.: A model curriculum for K-12 computer science: Report of the ACM K-12 Task Force Computer Science Curriculum Committee Ass. on for Computing Machinery, New York, NY (2003)
25. Gesellschaft für Informatik (GI) Grundsätze und Standards für die Informatik in der Schule, Bildungsstandards Informatik für die Sekundarstufe I. Addendum to LOG IN 28 (150/151) (2008), `http://www.informatikstandards.de/`
26. Kalas, I., Winczer, M.: Informatics as a Contribution to the Modern Constructivist Education. In: Mittermeir, R.T., Sysło, M.M. (eds.) ISSEP 2008. LNCS, vol. 5090, pp. 229–240. Springer, Heidelberg (2008)
27. Barber, M., Moursched, M.: How the world's best-performing school systems come out on top. McKinsey & Co. (2007)
28. Kalas, I.: Discovering Informatics Fundamentals through Interactive Interfaces for Learning. In: Mittermeir, R.T. (ed.) ISSEP 2006. LNCS, vol. 4226, pp. 13–24. Springer, Heidelberg (2006)
29. Dagiene, V.: Sustaining informatics education by contests. In: Hromkovič, J., Královič, R., Vahrenhold, J. (eds.) ISSEP 2010. LNCS, vol. 5941, pp. 1–12. Springer, Heidelberg (2010)
30. Dagiene, V.: The BEBRAS Contest on Informatics and Computer Literacy – Students (Drive to Science Education). In: Joint Open and Working IFIP Conf. Kuala Lumpur, pp. 214–223 (2008)

Why Teaching Informatics in Schools Is as Important as Teaching Mathematics and Natural Sciences*

Juraj Hromkovič and Björn Steffen

Department of Computer Science, ETH Zurich, Switzerland
{juraj.hromkovic,bjoern.steffen}@inf.ethz.ch

Abstract. In this paper, we aim to do more than arguing that informatics is a fascinating scientific discipline with interesting applications in almost all areas of everyday life. We pose the following questions: What are the educational requirements demanded from school subjects? Can we answer this question as satisfactory as we can do it, for instance, for mathematics, physics or chemistry? Does the teaching of informatics enrich education in ways other subjects cannot or do not sufficiently contribute?

Answering the questions above can not only be helpful for the discussion with politicians about integrating proper informatics and not only ICT skills in the educational systems, but it can also help us as teachers to focus on the fundamentals and on sustainable knowledge.

1 Introduction

If one deals with the problem of teaching a subject A in schools, one has to be able to answer the following three fundamental questions:

1. Does the teaching of A contribute to the understanding of our world and if yes, in which way and to what extent? How does A prepare the pupils for dealing with jobs and duties in society?
2. How important is the teaching of A for the ability to succeccfully study at a university? Do universities expect some fundamental knowledge of the discipline A?
3. What is the contribution of teaching A to the development of the way of thinking and the ability to solve various kinds of tasks and problems.

In the following three sections, we aim to give at least partial answers to these questions regarding the subject of informatics.

Before we deal with the three questions posed above, let us explain why we do not deal with imparting ICT skills in this article. Clearly, ICT skills have become important not only for studying, but also for future employments. Therefore, these skills need to be taught at school. However, in this article we focus on the unique, sustainable and lasting knowledge that informatics can provide.

---

* Partially supported by the Hasler Foundation.

I. Kalaš and R.T. Mittermeir (Eds.): ISSEP 2011, LNCS 7013, pp. 21–30, 2011.

To present scientific disciplines as collections of discoveries and research results gives a false impression. It is even more misleading to understand science merely in terms of its applications in everyday life. What would the description of physics look like if it was written in terms of the commercial products made possible by applying physical laws? Almost everything created by people—from buildings to household equipment and electronic devices—is based on the knowledge of physical laws. However, nobody mistakes the manufacturer of TVs or of computers or even users of electronic devices for physicists. We clearly distinguish between the basic research in physics and the technical applications in electrical engineering or other technical sciences. With the exception of the computer, the use of specific devices has never been considered a science.

Why does the public opinion equate the proficiency to use specific software packages to computer science? Why is it that in some countries teaching computer science is restricted to imparting ICT skills, i.e., learning to work with a word processor or to search on the internet? What is the value of such education, when software systems essentially change every year? Is the relative complexity of the computer in comparison with other devices the only reason for this misunderstanding?

Surely, computers are so common that the number of computer users is comparable to the number of car drivers. But why then is driving a car not a subject in secondary schools? Mobile phones are becoming small, powerful computers. Do we consider introducing the subject of using mobile phones into school education? We hope that this will not be the case. We do not intend to answer these rhetorical questions; the only aim in posing them is to expose the extent of public misunderstanding about computer science. Let us only note that experience has shown that teaching the use of a computer does not necessarily justify a new, special subject in school.

There are two main points in the relation between teaching core informatics and imparting ICT skills. First of all, none of our three principal questions posed above can be answered satisfactory if one reduces informatics to a computer driving licence. Secondly, reducing teaching of informatics to educating the usage of concrete software systems (word processor, spreadsheet, etc.) is the main reason for the current very bad image of informatics. Following our experience and some investigation in countries that replaced the teaching of informatics by imparting ICT skills, the pupils do not consider informatics as a science, they find it boring and not challenging enough to choose it as a subject for the study at a university. There is no other way out than to teach computer science in such a way that it is at least as attractive and deep as other natural sciences and mathematics.

2   Understanding the World Around Us

Physics helps us to understand a lot about the world around us. What is matter and what is energy? What is movement, space, and time? To understand at least some fundamental discoveries and laws of physics is of principal importance for understanding our world. Another example is, for instance, biology. It tries to uncover what life is and how it evolves. Many of the principal goals of the natural sciences are fascinating and challenging. The deep questions about the functioning of our world and about our role in this world are for most young people much more attractive than any colorful application on the screen. Understanding nontrivial concepts and getting deep insight into complex subjects makes them more pleased than mastering any skill.

Can computer science provide something comparable? Our answer is yes. The following two discoveries of computer science are of this kind.

The limits of automation and the concept of algorithms

All possible tasks (problems) can be partitioned into two classes, namely algorithmically solvable and algorithmically unsolvable tasks. Due to the introduction of the notion of the algorithm by A. Turing [13], we got an instrument for classifying computing problems into those solvable by computers and others unsolvable by computers. We discovered the existence of a rigorously defined limit of automation.

Quantitative laws of information processing and the concept of computational complexity

To perform a computation or to process information in order to solve a concrete task requires some amount of work. There exist quantitative laws of information processing. For each task of information processing, one can investigate the amount of computer work sufficient and necessary in order to solve the task. We discovered that more computational resources help us to solve more tasks and that there are arbitrarily hard tasks with respect to the amount of work required. Thousands of practical computing problems cannot be solved because the amount of work necessary is beyond the physical capabilities of our universe. The core scientific topic of informatics is to recognize how much information one can extract from the given data by a reasonable amount of work.

Both fundamental concepts mentioned above do not only reveal us the existence of some natural laws of information processing. Similar to physics, these concepts are also crucial for many of today's applications. For instance, the current e-commerce, based on public-key cryptography, would not exist without the fundamental concepts of algorithms and computational complexity.

Examples of how to introduce these concepts in an appropriate way for secondary schools is presented in numerous books: *Algorithmics: The Spirit of Computing* [7] *Das Affenpuzzle* [3] *Abenteuer Informatik* [6], *Computer Science Unplugged* [1], *Taschenbuch der Algorithmen* [14], *Algorithmic Adventures* [10], *Lehrbuch Informatik* [8], *Berechenbarkeit* [9], *Einführung in die Kryptologie* [5].

Another important issue about computational complexity is the fact that many computing problems can be solved efficiently, but it is a challenge to discover an efficient algorithm for them and this subject of informatics is very fruitful.

A classical problem from cryptology that is not practically solvable using a naive approach is for example the calculation of powers with very large exponents. But with an ingenious trick, the so-called fast exponentiation, this is efficiently doable. In Appendix A we present an approach we used in the textbook *Einführung in die Kryptologie* [5] to introduce the fast modular exponentiation to secondary-school students.

We conclude that informatics has a nontrivial scientific depth that is fascinating and can challenge young people with its goals and problems. Informatics expands science also by giving new dimensions to the fundamental categories of science such as determinism, nondeterminism, randomness, proof, simulation, correctness, efficiency etc. Many of its discoveries are surprising and attractive to young people. For instance, exchanging a deterministic control by a randomized one, whose decisions are partially influenced by random bits, can essentially decrease the amount of work necessary to reach the intended goals.

We only need to get more experience with teaching these topics understandably in our schools. There is no doubt that, in our society based on knowledge which is extracted daily from a huge amount of data, one cannot understand the world around us without some fundamental knowledge of informatics. The same is true for the control of technical devices in the technical world created by man.

3   Preparation for University Studies

For sure, ICT skills have become instruments that necessarily have to be mastered to successfully participate in many human activities. But this is not the matter of our discussion because they are mostly on the basic technical level like using a pen for writing. Here we want to look for the usefulness of sustainable knowledge in informatics.

First of all, most of the scientific disciplines are asked to handle a huge amount of data. This is not only true for experimental sciences such as biology, chemistry or physics, but also for economics, sociology, medicine etc. Because the amount of data is growing fast, one needs a well-structured way of saving it as well as efficient algorithms for searching, processing and communicating it. Many scientific disciplines, not only the natural sciences, generate knowledge from the data they collected by experiments, measurements, and assessments. To extract knowledge from the given data often requires a huge amount of computer work. This can only be done if one is able to develop specific efficient algorithms for these tasks or to ask computer scientists for their support. But this cannot be achieved without the ability to describe the problems exactly.

Of a similar importance is the application of simulations. In order to forecast some development, we create models and run simulations on them. This originally basic research instrument of physics, chemistry and engineering became a fundamental tool in sciences like economics, sociology, psychology, etc., which avoided the use of exact mathematical methods for a long period of time. Without simulations, many research projects would be unthinkable.

Nobody discusses whether mathematics has to be a part of school education or not. But informatics is the scientific discipline making mathematics to a technology. Due to this, mathematics is applied everywhere.

Programming is a part of informatics with a growing importance. All technical systems are controlled by programs and therefore all students of technical sciences need to master programming. Slowly, but surely, this becomes true also for several areas of the natural sciences and even for humanities. Programming is not only the ability to implement given methods into programs. Much more important is the ability to use abstractions to describe problems, to analyse them and to find methods for solving them.

Definitely, we conclude that teaching informatics in school is not only a contribution for the study of related topics at university. The knowledge about the capabilities and limits of computer science and the way of working in informatics is at least as useful for the study at a university as any other classical subject of high school curricula.

4   A Way of Thinking and Working

One of the main arguments for teaching mathematics is the development of the exact way of thinking that finally results in the ability to use the exact language of mathematics for describing, analyzing and solving problems in all areas of our life. This ability becomes more and more important. Some colleagues tend to call informatics the "new" mathematics or at least a constructive mathematics. Jeannette Wing, head of the computer science department at Carnegie Mellon University, even envisions that "thinking like a computer scientist" should be a fundamental skill such as reading, writing and arithmetics [15]. Cohen and Haberman regard computer science as one of the five "languages" every citizen should acquire [2].

The reason for that is the way of working in computer science. Similarly as in mathematics, we begin with an abstract description of a problem and continue with its analysis. But additionally, computer scientists do not only discover an efficient way of solving it, but they also implement the discovered method and provide a product (program) for solving problems of this kind. This work is more constructive than the typical work of a mathematician and ties the exact way of thinking in mathematics with the pragmatic way of working in engineering.

In this way teaching informatics in school:

- supports the development of the exact way of thinking and working in mathematics and natural sciences, and
- brings new elements to education by teaching elements related to the way of thinking and working in engineering disciplines (introducing the concepts of implementation, verification (proving correctness), testing, modular design, etc.).
- supports interdisciplinarity, because the computing task considered (the extraction of information given data) have their origin in various scientific disciplines and industrial applications.

No other subject in school goes this long way from a problem formulation to a solution in the form of a product (program). This high level of constructivism is probably the main contribution of teaching computer science. This approach also contributes to the teaching of math as Syslo and Kwiatkowska point out [12].

It has to be said that teaching informatics in this way is very rewarding. One can fascinate young people and give them the great feeling of achievement.

Instead of being first of all an examinator, checking the success of the class by exams, the teacher can switch into the role of a supporter helping along the way from the problem to the solution. Whether the result of their work is correct can be directly verified by the pupils, no immediate judgement of the teacher is necessary. Additionally, team work can be educated in an excellent way.

5   Conclusion

To teach the discoveries of informatics, its methods, and ways of thinking and acting contributes to education at least in the same amount as teaching mathematics or other classical subjects. Additionally, teaching informatics brings new elements to the schools that we are missing already for a long time. Informatics contributes to science with new fundamental concepts and terms like algorithm, computational complexity, efficiency, verification, simulation, information security, etc., and gives a deeper insight on some fundamental notions of science such as determinism, nondeterminism, and randomness, to name a few.

Teaching informatics has became important to successfully study at a university in many scientific disciplines. More and more scientific disciplines expect and will expect fundamental knowledge of informatics and especially the ability to apply it in their own discipline.

The way of thinking and working in informatics enriches the human way of thinking and can essentially contribute to the success of young people in their life, whatever they will do in the future.

References

1. Bell, T., Fellows, M., Witten, I.H.: Computer Science Unplugged - Off-line activities and games for all ages (2006), http://www.csunplugged.org
2. Cohen, A., Haberman, B.: Chamsa: Five languages citizens of an increasingly technological world should acquire. ACM Inroads 1, 54–57 (2010)
3. Davis, H.: Das Affenpuzzle. Springer, Heidelberg (2001)
4. Diffie, W., Hellman, M.: New directions in cryptography. IEEE Transactions of Information Theory 22(6), 644–654 (1976)
5. Freiermuth, K., Hromkovič, J., Keller, L., Steffen, B.: Einführung in die Kryptologie. Vieweg+Teubner (2009)
6. Gallenbacher, J.: Abenteuer Informatik, 2nd edn. Elsevier, Amsterdam (2008)
7. Harel, D., Feldman, Y.: Algorithmics: The Spirit of Computing, 3rd edn. Addison Wesley, Reading (2004)
8. Hromkovič, J.: Lehrbuch Informatik. Vieweg+Teubner (2008)
9. Hromkovič, J.: Berechenbarkeit. Vieweg+Teubner (2011)

10. Hromkovič, J.: Algorithmic Adventures. Springer, Heidelberg (2009)
11. Rivest, R.L., Shamir, A., Adleman, L.M.: A method for obtaining digital signatures and public-key cryptosystems. Commun. ACM 21(2), 120–126 (1978)
12. Syslo, M., Kwiatkowska, A.: Contribution of informatics education to mathematics education in schools. In: Mittermeir, R.T. (ed.) ISSEP 2006. LNCS, vol. 4226, pp. 209–219. Springer, Heidelberg (2006)
13. Turing, A.M.: On computable numbers with an application to the Entscheidungs-problem. Proceedings of London Mathematical Society 42(2), 230–265 (1936)
14. Vöcking, B., Alt, H., Dietzfelbinger, M., Reischuk, R., Scheideler, C., Vollmer, H., Wagner, D. (eds.): Taschenbuch der Algorithmen. eXamen.press, Springer (2008)
15. Wing, J.: Computational thinking. Comunications of the ACM 49(3) (2006)

A   Fast Modular Exponentiation

Many cryptographic protocols, e. g. the RSA cipher [11] and the Diffie-Hellman protocol [4], rely on the calculation of modular powers with very large exponents, i. e. numbers with hundred or more digits. This is one prime example where the naive approach to calculate it leads to an inefficient algorithm, but with a important observation we can determine these powers efficiently.

This section shows an excerpt from our textbook *Einführung in die Kryptologie* [5], where we explain the method of the fast modular exponentiation to secondary school students.

If we calculate the power $a^x \bmod n$ naively like

$$a^x \bmod n = \underbrace{a \cdot a \cdot a \cdot \ldots \cdot a}_{x \text{ factors}} \quad \bmod n$$

with repeated multiplications, then $x - 1$ multiplications have to be carried out.

When x is large, 10^{200} for example, then more multiplications have to be made than the age of the universe ($\approx 10^{17}$ seconds) multiplied by the number of particles in the visible universe (below 10^{80}). This means that the calculation of a power like $a^{10^{200}}$ is not feasible, if we carry it out so clumsy. Therefore we want to build a more clever algorithm to efficiently calculate such powers.

First of all we observe that we do not need to work with large numbers such as a^x because we work modulo n and the students already know that they are allowed to reduce the size of numbers in the calculation below n by computing the reminder modulo n after each computation step[1]. Therefore, the size of the representation of the numbers in our calculation is always in $O(\log n)$ and we can measure the computational complexity (amount of computer work) by the number of arithmetic operations executed.

Therefore, we are allowed to focus on the number of operations needed to compute a^x and so we simplify our exponentiation by removing modular calculations mod n from our notation.

---

[1] At this point the students already worked trough a module about modular computations in the textbook [5].

We now ask the students to compute a^x with fewer than $x - 1$ multiplications for concrete numbers for x. They may discover it by their own or you can present a few motivating examples such as the following ones:

The power a^{16} can be expressed as

$$a^{16} = \left(\left(\left(a^2\right)^2\right)^2\right)^2$$

and because of that one can calculate the power as follows:

$$a^2 = a \cdot a$$
$$a^4 = a^2 \cdot a^2$$
$$a^8 = a^4 \cdot a^4$$
$$a^{16} = a^8 \cdot a^8$$

by using only 4 multiplications. Similarly the following power

$$a^{24} = \left(\left(\left(a^3\right)^2\right)^2\right)^2$$

can be determined with only 5 multiplications:

$$a^2 = a \cdot a$$
$$a^3 = a^2 \cdot a$$
$$a^6 = a^3 \cdot a^3$$
$$a^{12} = a^6 \cdot a^6$$
$$a^{24} = a^{12} \cdot a^{12}.$$

Then plenty of small challenges can be formulated. For instance, we can ask for

- calculating a^x for a concrete x with a given number of multiplications.
- searching for the smallest number of multiplications sufficient to compute a^x for concrete values of x.
- finding several different optimal ways for calculating a^x for concrete values of x.

After playing with several small challenges of the above mentioned kind, one can pose the following question:

"It is nice to discover the best possible ways for calculating a^x for a given x. But this does not provide the possibility to automize the calculation, because for different x and y we may use different approaches to calculate a^x and a^y. Hence, we need an efficient algorithm that calculates a^x for any x in an uniform way."

One can start to search for some systematic way of calculating a^x for any x. With some help the class may be able to discover it on their own. A good starting point is an example of the following kind:

$$a^{21} = a^{16} \cdot a^4 \cdot a = \left(\left(\left(a^2\right)^2\right)^2\right)^2 \cdot \left(a^2\right)^2 \cdot a.$$

We can calculate a^{21} by 6 multiplications as follows:

$$a^2 = a \cdot a$$
$$a^4 = a^2 \cdot a^2$$
$$a^8 = a^4 \cdot a^4$$
$$a^{16} = a^8 \cdot a^8$$
$$a^{20} = a^{16} \cdot a^4$$
$$a^{21} = a^{20} \cdot a.$$

We observe that this calculation has two parts. The first one is used to compute a^2, a^4, a^8, and a^{16}, and the second one calculates the product of some of them. In this way one can discover that the binary representation of 21 is

$$10101,$$

where the first 1 is for 16 (2^4), the second 1 for 4 (2^2) and the last 1 stands for 1 (2^0). This means

$$a^{21} = a^{16} \cdot a^4 \cdot a,$$

i.e., the binary representation estimates, which of the partial precomputed powers a, a^2, ..., a^{16} are used to compute the final result a^{21}. In general it is true that if

$$x = x_n \cdot 2^n + x_{n-1} \cdot 2^{n-1} + \cdots + x_1 \cdot 2^1 + x_0 \cdot 2^0 = \sum_{i=0}^{n} x_i \cdot 2^i,$$

then

$$a^x = a^{x_n \cdot 2^n} \cdot a^{x_{n-1} \cdot 2^{n-1}} \cdot \ldots \cdot a^{x_1 \cdot 2} \cdot a^{x_0} = \prod_{i=0}^{n} a^{x_i \cdot 2^i}.$$

The analysis of the computational complexity of this algorithm can be estimated easily. If $x = x_n x_{n-1} \ldots x_2 x_1 x_0$ is the binary representation of x, then we need first n multiplications to compute a^2, ..., a^{2^n} and then as many multiplications as there are 1's in the binary representation of x. Since $\lfloor \log_2(x) \rfloor \leq n$,

$$2 \cdot \lfloor \log_2(x) \rfloor$$

multiplications are always sufficient to calculate a^x.

That does not need to be the end of our short teaching unit. One can deepen the acquired knowledge by posing and investigating for instance the following questions:

1. An algorithm for computing a^x can start as follows:

$$a^2 = a \cdot a$$
$$a^3 = a^2 \cdot a$$
$$a^9 = a^3 \cdot a^3 \cdot a^3$$
$$a^{27} = a^9 \cdot a^9 \cdot a^9$$
$$a^{81} = a^{27} \cdot a^{27} \cdot a^{27}$$
$$a^{3^n} = a^{3^{n-1}} \cdot a^{3^{n-1}} \cdot a^{3^{n-1}}$$

for $3^n \leq x < 3^{n-1}$. Can you complete the description of this algorithm in order to compute a^x and estimate its computational complexity? Can this algorithm be better or equally efficient compared to the algorithm based on the binary representation of x?

2. Is it profitable to execute the first part of the algorithm from question 1 as follows?

$$a^2 = a \cdot a, \qquad\qquad a^3 = a^2 \cdot a,$$
$$a^6 = a^3 \cdot a^3, \qquad\qquad a^9 = a^6 \cdot a^3,$$
$$a^{18} = a^9 \cdot a^9, \qquad\qquad a^{27} = a^{18} \cdot a^9,$$

etc.

Or asked differently, is it helpful to additionally compute and store the powers a^6, a^{18}, etc., as well?

3. Develop a new algorithm for the computation of a^x based on the 5-ary representation of x. How well does this algorithm behave compared to the previous ones?

4. Does the "binary representation" algorithm always (for all x) work in the most efficient way or do there exist values for x, for which one can compute a^x faster?

Informatics Education in Italian High Schools

Maria Carla Calzarossa[1], Paolo Ciancarini[2],
Luisa Mich[3], and Nello Scarabottolo[4]

[1] University of Pavia, Italy
mcc@unipv.it
[2] University of Bologna, Italy
ciancarini@cs.unibo.it
[3] University of Trento, Italy
luisa.mich@unitn.it
[4] University of Milano, Italy
nello.scarabottolo@unimi.it

Abstract. This paper presents the main results of an extensive monitoring exercise aimed at assessing the role of informatics education in Italian high schools. The investigation focused on the teaching and certification activities performed by the schools as well as on the role and use of information technologies for teaching and communications with the students and their families. The study has shown a very positive attitude of the schools towards informatics education: many offer specific courses to their students and promote the use of the technologies for teaching a large variety of disciplines. The certification process of ICT skills is also popular. Despite all these positive aspects, the investigation has shown that informatics education is often limited to the use of computers and seldom addresses the foundations of the discipline.

Keywords: Informatics education, high schools, ICT certifications.

1   Introduction

Information and Communication Technologies (ICT) play a crucial role for our daily personal and professional activities. The worldwide diffusion of online social networks among young people – that is, the so-called "digital generation"– is a proof of their familiarity with these technologies, typically used straightaway with little or no difficulty at all [10]. Nevertheless, the knowledge of the digital generation of the foundations and principles upon which ICT works is usually rather superficial and often not properly included and blended in their education (see for instance [1, 4, 12]). Education systems should then take the responsibility to fill this gap by implementing some specific learning processes.

Another important step in these processes is represented by independent certification exams aimed at assessing the knowledge and skills acquired. The relevance of ICT certifications in the job market is confirmed by the large number of organizations investing in this field. Among the most renowned organizations, we can name CompTIA (http://www.comptia.org), an international organization whose mission is focused on ICT, and AQA (http://www.aqa.org.uk),

I. Kalaš and R.T. Mittermeir (Eds.): ISSEP 2011, LNCS 7013, pp. 31–42, 2011.
© Springer-Verlag Berlin Heidelberg 2011

that provides ICT certifications in a broader range of domains and educational initiatives.

The impact of ICT certifications in Italian Universities is addressed in [2], whereas very few papers specifically analyze the use of certifications in schools. For example, in [11] the values conveyed by offering ICT certifications at the high school level are illustrated, while in [6] the rationale for the diffusion in the schools of a specific proprietary certification is discussed.

On the contrary, the analysis of informatics education in high schools and the impact of ICT in the overall education system have been extensively addressed. Comparative studies of the educational approaches adopted in various countries are presented in [7]. In [13], authors outline that even though colleges and universities emphasize the importance of ICT for their students, the ICT skill level for incoming freshmen is often below the standard required for academic success. Nevertheless, students believe they have excellent technological skills, possibly because they compare their skills to those of their parents or of high school teachers belonging to older generations. The situation in Austrian academic secondary schools is presented in [8]. In particular, the analysis has shown that different autonomous approaches adopted by schools in the implementation of their informatics education have led to some undesirable digital gaps at the end of the lower secondary level. As a consequence, a more ambitious informatics education becomes infeasible in academic secondary schools. An analysis of the Lithuanian situation is described in [5], where the author discusses the goals for introducing ICT in high schools and the competencies and values to be developed for informatics education. In [9], the authors focused on the answers of about 400 Israeli high school students who were asked about their ICT skills and knowledge.

A recent report [14] has shown that in the USA roughly two-thirds of the states have few computer science education standards for secondary school education and most states treat high school computer science courses as simply an elective and not as part of a student's core education. Moreover, much of what passes for high school computer science instruction is actually about Information Technology literacy rather than algorithm design, programming, or computational thinking.

This paper presents the results of an extensive monitoring exercise performed in the year 2010 and aimed at assessing how informatics education is perceived and organized in Italian high schools. The investigation, that involved the schools of eight Italian Regions, focused on various aspects related to teaching and certification activities performed by the schools as well as on the role of information technologies for teaching other disciplines. A previous investigation, performed in the year 2008, whose main outcomes are reported in [3], considered a much smaller number of Regions and schools and covered some of these aspects to a more limited extent.

The paper is organized as follows. Section 2 introduces the methodological approach applied for the investigation and describes the main characteristics of the schools involved in the exercise. The positions of the schools with respect to ICT are illustrated in Section 3, whereas the organization adopted by the schools to teach informatics is described in Section 4. Section 5 presents the projects of the schools in the framework of informatics certification. Finally, some concluding remarks are outlined in Section 6.

2   Methodological Approach

2.1   The Context

Italian high schools extend over five years. Within the Italian high school system, informatics disciplines are not considered as compulsory disciplines for every type of school. In some schools – mainly technical schools – they are part of their curricular programmes, in some others these disciplines are often part of their so-called extra-curricular activities, whose organization and contents are under the responsibility of the individual schools. Some schools do not offer any informatics course at all. It is worth to point out that the Italian high school system has undergone a recent reform whose outcomes will be visible in about five years time. Table 1 presents as an example the breakdown of the teaching hours foreseen by the reform for various disciplines for technical schools addressing the ICT specialization.

Table 1. Breakdown over five years of the teaching hours in technical schools specialized in ICT based on the 2010 Italian high school reform

Scientific and technical disciplines	1	2	3	4	5
Physics	99	99			
Chemistry	99	99			
Graphic design	99	99			
Informatics	99				
Applied sciences and technologies		99			
Complements of mathematics			33	33	
Systems and networks			132	132	132
Design of ICT systems			99	99	132
Project and enterprise management					99
For Informatics curriculum					
Informatics			198	198	198
Telecommunications			99	99	
For Telecommunication curriculum					
Informatics			99	99	
Telecommunications			198	198	198
Yearly total	396	396	561	561	561
Yearly total (including other disciplines)	1056	1056	1056	1056	1056

2.2   The Monitoring Exercise

The monitoring exercise relies on a web-based questionnaire designed by the authors and sent to the Directors of the high schools of eight of the 20 Italian Regions, that is, Apulia, Lazio, Lombardy, Marche, Molise, Sicily, Umbria and Veneto. The choice of these Regions was mainly dictated by their number and type of schools: some of these Regions, such as, Lombardy, Apulia and Sicily, are characterized by the largest number of schools and student population in Italy, whereas others, such as, Molise and Umbria, have very few schools and students. Moreover, the geographical location of these eight Regions provides a good and representative coverage of the entire country from North to South, thus, taking into account the different socio-economics settings.

The investigation was launched in the spring 2010 and focused on the classes of the final three years. In total some 1,220 schools, out of 2,776 invited to participate, responded to the questionnaire, that is, approximately, 44%. From the regional distribution of the schools (see Fig. 1), we can notice that about one-fourth of the schools are located in Lombardy and slightly less than 20% in Sicily, whereas only 1.9% are located in Molise.

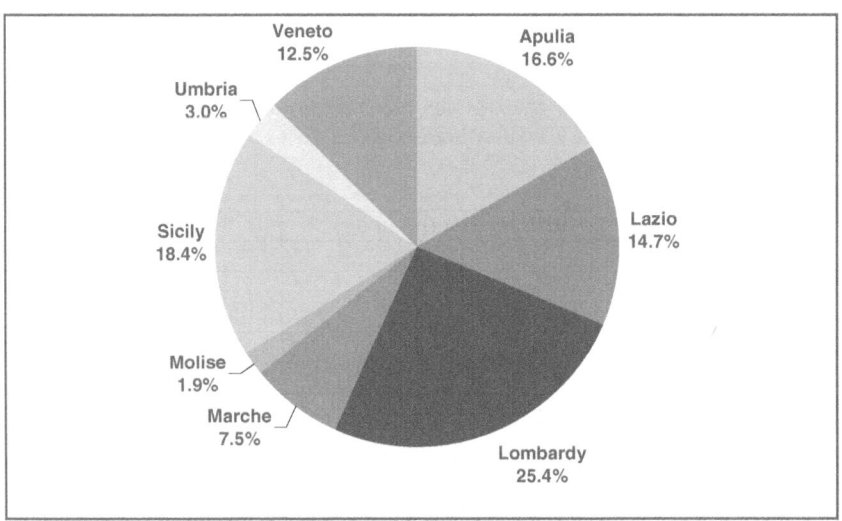

Fig. 1. Regional distribution of the schools

The total number of students enrolled in the final years of these schools is approximately 356,000, that is, slightly less than 300 students per school and some 19.5 students per class. About one-third of these students, namely, 127,000, are enrolled in technical schools.

3   ICT in the Italian High Schools

To analyze the role of ICT in the schools, our monitoring exercise first focused on the availability of PCs for the teaching activities. The investigation has shown that PCs are available in about 96% of the schools, with a total of some 78,500 PCs. These PCs are used as teaching tools for a large variety of disciplines. As can be seen from Figure 2, very many schools employ PCs for teaching mathematics (82% of the schools) and foreign languages (76% of the schools), whereas PCs are used only by about half of the schools for teaching drawing. Among the other

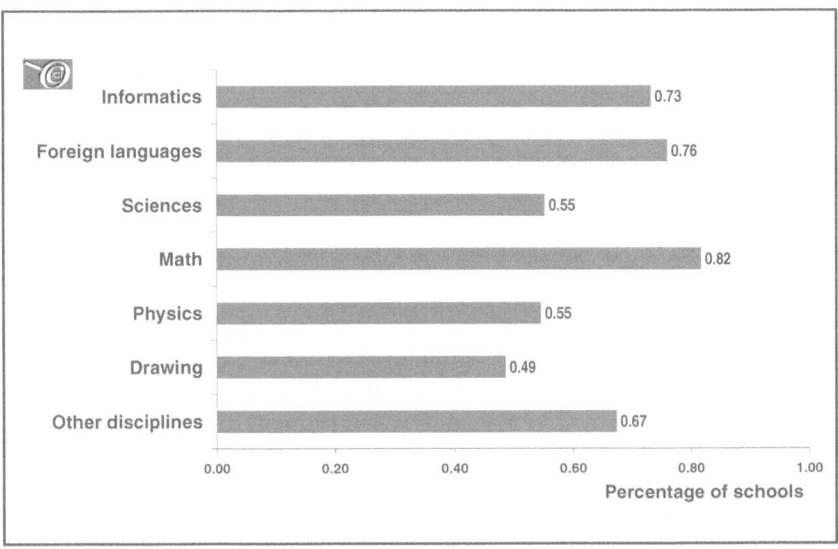

Fig. 2. Disciplines taught with the support of PCs

disciplines selected by the schools, the most popular are humanities, such as, Italian, Latin and ancient Greek, and technical disciplines, such as, aeronautics, agronomy, mechanical constructions, advertising technologies.

The technical support for the management of PCs and ICT infrastructures (e.g., installation, configuration, update) is provided by staff specifically employed by the schools for this purpose. On average, each school employs 3.6 persons. However, the variability of the distribution among schools is rather large. For example, in some technical schools this staff consists of more than 20 persons, whereas other schools do not employ any dedicated staff at all and their duties are often performed by the teachers on a voluntary basis.

The majority of the schools, i.e., 71%, is also equipped with a large variety of technologies: servers, routers, wired and wireless networks, multimedia

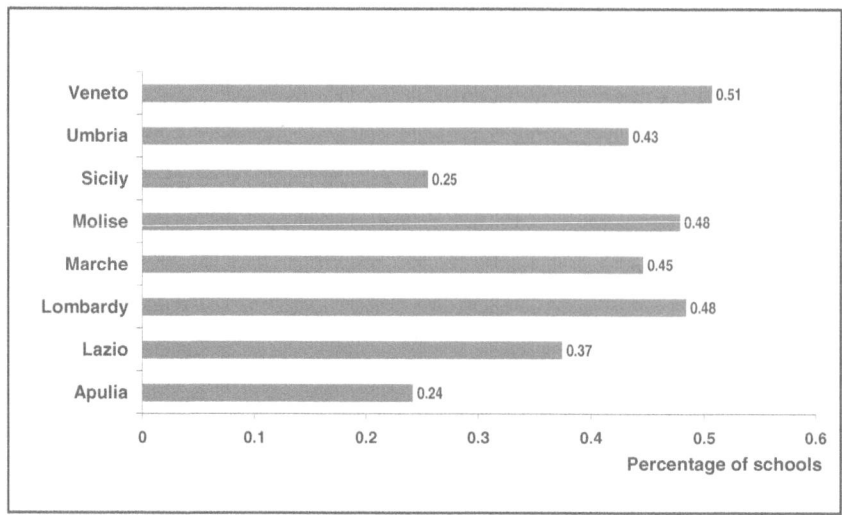

Fig. 3. Use of email within the schools of the eight Regions

interactive whiteboards, overhead projectors, camcorders, audio mixers. It is important to outline that a good number of schools, i.e., 600 schools, adopts specific technologies to help students with disabilities in their learning process.

Email is rather popular within schools, even though, as Figure 3 shows, there are large differences among the schools of the various Regions.

With respect to the exploitation of Web technologies, more than 92% of the schools have their own Web site usually maintained and updated by the teachers of the schools. We did not notice any significant difference among the schools of the various Regions. Schools often post on their Web sites confidential information about their student careers, e.g., absence, grades, overall assessments. For privacy reasons – in Italy privacy regulations are very strict – this sensitive information is only accessible by the parents, via login and password.

Despite the popularity of Web sites, very few schools (i.e., 192) have adopted advanced Web technologies, such as, online social networks and forums, involving teachers and students. The schools which approached these technological solutions are located in two Regions only: Apulia and Veneto.

Similarly, the organization of entertainment activities based on technological competence, such as videogames contests and digital art, is very limited. On the contrary, several schools are very active in participating to various types of international contests, such as the International Olympics in Informatics.

4   Teaching Organization

Our investigation has shown that informatics disciplines are taught in about 743 schools (out of the 1,220 that responded to our questionnaire). We remark that according to the current Italian high school system, schools can teach these disciplines in mandatory courses taken by all students of a class and in optional courses taken by their students on a voluntary basis.

In particular, we can subdivide the courses offered by the schools in four different categories:

- mandatory courses belonging to the basic curriculum of a class;
- additional mandatory courses not belonging to the basic curriculum of a class, whose offer is left up to individual schools;
- optional free courses;
- optional courses requiring the payment of a fee.

Figure 4 shows the distribution of the schools according to the types of courses offered to their students. As can be seen, there is the prevalence of mandatory courses of informatics belonging to the basic curriculum of a class (offered by 441 schools, this is, 59.4% of the schools where informatics is taught). A good number of schools, i.e., 274, offer optional free courses of informatics. Of course each school can organize courses in each of these categories. In particular, slightly less than 300 of the schools that participated to our monitoring exercise offer both mandatory and optional courses. A more detailed investigation has shown that most of these schools are technical schools. This demonstrates their strong attitude and interest towards informatics disciplines that are considered as a compelling professional need for the future of their students.

To further explore the choices of the schools, we focused on the organization of mandatory courses and in particular on the projects developed by the students

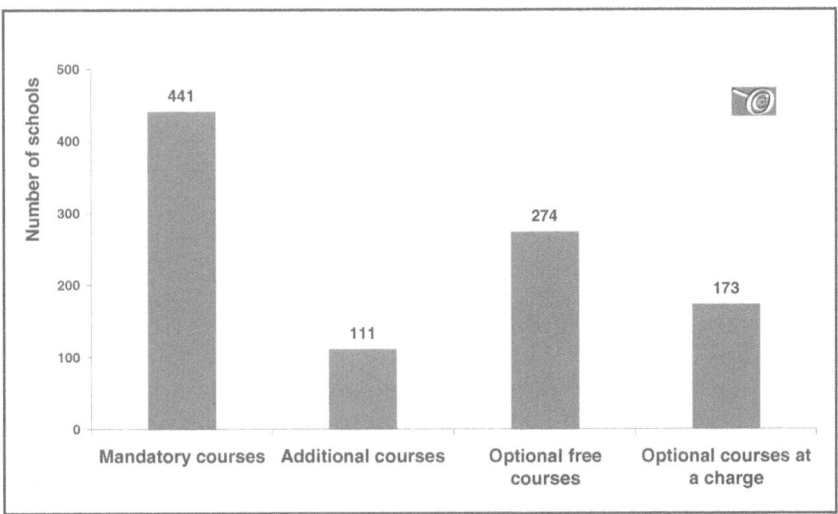

Fig. 4. Schools which teach informatics disciplines

as assignments. The investigation has shown that this type of assignment is adopted by almost 60% of the schools. As can be seen from Figure 5, most projects involved some programming activities. Very popular are also projects dealing with the development of Web sites, whereas far less popular are those addressing hardware devices.

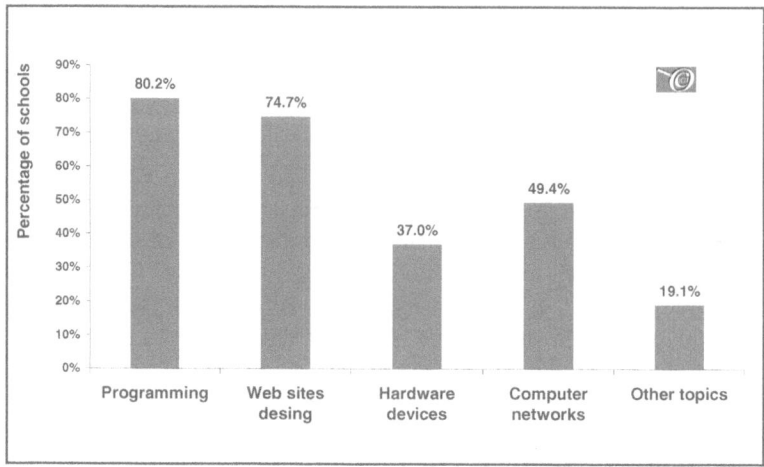

Fig. 5. Topics addressed by the projects developed in the framework of mandatory courses

Despite mandatory courses, whose topics are defined by national regulations, the organization and content of optional courses are up to the schools. It is interesting to outline that very many optional courses focused on the use of individual productivity software suites, such as, Microsoft Office suite. Other topics, such as, Internet and navigation tools, basic concepts of informatics, are often part of these courses. Very few courses are dedicated to programming languages.

In general, it is possible to observe a bias towards teaching informatics from the end-user perspective, with a particular focus on productivity tools and Web technicalities and very seldom concentrating on fundamental aspects, such as, algorithmic aspects, logic, programming.

The average number of hours taught for each of the topics offered in the framework of optional free courses of informatics is shown in Figure 6. We can observe the large number of hours dedicated to teach programming languages.

Note that about half of the schools that responded to our questionnaire organize courses on ICT for their teachers – a rather unusual practice in a country where competence upgrade and lifelong learning of teachers are not planned nor regulated. These courses usually cover a large variety of topics, ranging from teaching technologies and the use of multimedia interactive whiteboards to the use of the Microsoft Office suite.

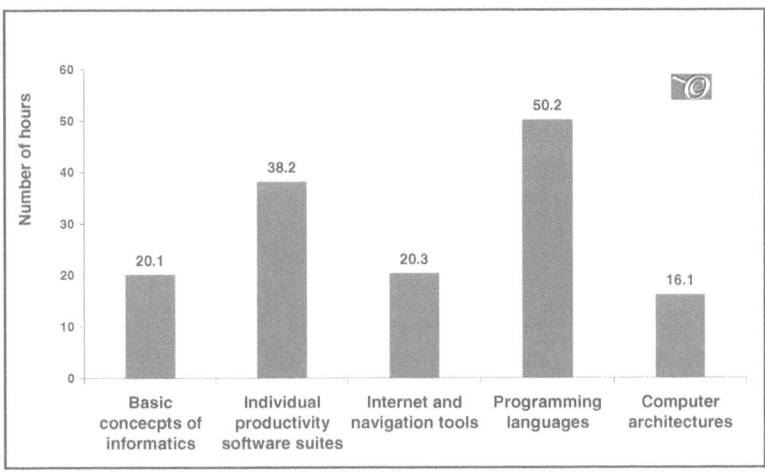

Fig. 6. Hours for the various topics addressed within the optional free courses offered by the schools

5 Certification of ICT Skills and Education

More than half of the schools (631 out of the 1,220 that participated to our investigation) offer certification exams to their students. This approach for establishing a quality standard is especially popular in the schools of Veneto, Marche and Apulia, slightly less in Umbria and Sicily. The diffusion of certification programs is comparatively much less popular in the high schools of Lazio. Thus, regional differences exist but there is no clear explanation why.

Most high schools offer the ECDL certification (European Computer Driving Licence), with the majority of projects based on the ECDL FULL certification, obtained by passing all the seven tests foreseen by the programme, with respect to ECDL START certification, obtained by passing four out of the seven tests. In detail, the total number of certifications obtained by the students during the school year 2008-2009 was equal to 24,000. It is worth noting that 84% of these certifications refer to the ECDL family. In particular, 16,000 students received the ECDL FULL certification, whereas some 4,500 the START one. Moreover, about 1,000 students received a CISCO certification, and about 800 a Microsoft certification.

Let us remark that almost 80% of the schools that offer ICT certifications organize the certification exams within the school itself. This is mainly the case of the schools of Lombardy, Molise and Umbria.

Concerning the costs related to the certification process, we have noticed (see Fig. 7) that for the majority of the schools these costs are fully or partly covered by the families of the students. Very few schools provide their students with some financial support based on their grades or on the income of their family.

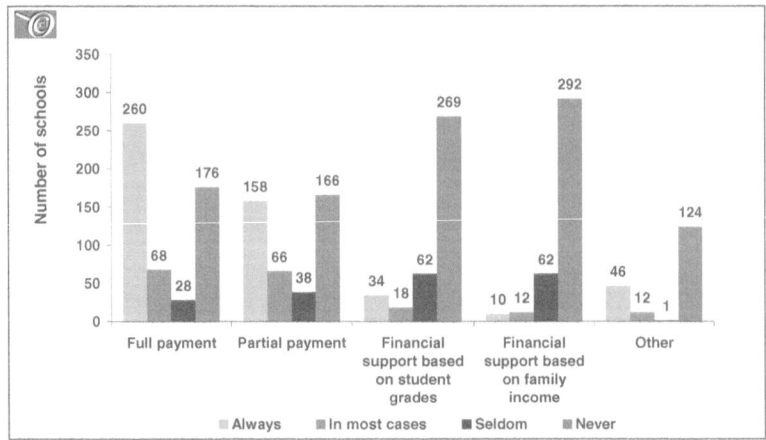

Fig. 7. Behavior of the schools with respect to the coverage of the costs associated with the ECDL certification

It is interesting to outline that about 30% of the schools that offer ICT certifications complement them with other types of certifications, such as, language certifications for testing English proficiency. Moreover, about 10% of the schools are interested to widen their offers. These data witness a genuine interest of the schools to audit and validate the knowledge and skills developed by their students.

6   Conclusions

The extensive monitoring exercise carried out in the Italian high schools has shown their general interest towards informatics education: most schools offer several courses, either mandatory or optional, that covers various aspects of the discipline and involve a good number of students. Similarly, ICT certifications are strongly promoted by the schools that are usually equipped for an in-house implementation of the certification tests. It is interesting to note that these certifications very often complement English language certifications, thus, showing a general, positive attitude of Italian high schools with respect to third-party independent assessment of students skills.

Technologies play an important role in the schools: a very large number of schools is equipped with a big variety of technologies that are used to teach many different disciplines. Web sites and email are also very popular, much less are other Web technologies.

Despite all the positive aspects outlined above, let us remark – as already mentioned in Sect. 4 – that the approach adopted by Italian high schools for teaching informatics is not always in line with their mission. In our digital age, schools should provide their students with fundamental problem solving knowledge and skills required for living and working in our society and problem solving by computers requires some specific education. However, in most of the schools informatics courses address simply the use of the computers, and only very few courses focus on the foundations of informatics, such as, computer architecture, programming, computational thinking. This situation has two negative implications: there is a risk for students to perceive informatics as a discipline with no scientific challenges but barely a matter of shallow playing with technologies. Moreover, the best high school students very rarely enroll in informatics degrees at the university, thus hindering the technological advancements in this field.

A different teaching framework and strategy for informatics education should then be defined as to provide the students with the skills required by next generation technologies. As a future work, we plan to analyze in details the strategies of the other European countries towards informatics education.

Acknowledgments. Authors gratefully thank AICA, CINI and Fondazione CRUI for their encouragement and continuous support during all phases of this monitoring exercise.

References

1. Brinda, T., Puhlmann, H., Schulte, C.: Bridging ICT and CS: educational standards for Computer Science in lower secondary education. ACM SIGCSE Bull. 41(3), 288–292 (2009)
2. Calzarossa, M., Ciancarini, P., Maresca, P., Mich, L., Scarabottolo, N.: The ECDL Programme in Italian Universities. Computers & Education 49(2), 514–529 (2007)
3. Calzarossa, M., Ciancarini, P., Mich, L., Scarabottolo, N.: ICT teaching and certification in Italian high schools. In: Hermann, C., Lauere, T., Ottmannn, T., Welte, M. (eds.) Informatics Education Europe IV - IEE IV, pp. 89–94. University of Freiburg (2009)
4. Centre for Educational Research and Innovation: New Millennium Learners. Initial findings on the effects of digital technologies on school-age learners. In: OECD/CERI International Conference on Learning in the 21st Century: Research, Innovation and Policy (2008),
http://www.oecd.org/dataoecd/39/51/40554230.pdf
5. Dagiene, V.: Teaching information technology and elements of informatics in lower secondary schools: Curricula, didactic provision and implementation. In: Mittermeir, R.T., Sysło, M.M. (eds.) ISSEP 2008. LNCS, vol. 5090, pp. 293–304. Springer, Heidelberg (2008)
6. Dennis, A., Duffy, T., Cakir, H.: IT programs in high schools: lessons from the Cisco Networking Academy program. Communications of the ACM 53(7), 138–141 (2010)
7. Law, N., Pelgrum, W., Plomp, T. (eds.): Pedagogy and ICT use in schools around the world - Findings from the IEA sites 2006 study. CERC Studies in Comparative Education, vol. 23. Springer, Heidelberg (2008)
8. Micheuz, P.: Some findings on Informatics education in Austrian academic secondary schools. Informatics in Education 7(2), 221–236 (2008)
9. Nachmias, R., Mioduser, D., Shemla, A.: Information and Communication Technologies usage by students in an Israeli high school: Equity, gender, and inside/outside school learning issues. Education and Information Technologies 6(1), 43–53 (2001)

10. Palfrey, J., Gasser, U.: Born digital: understanding the first generation of digital natives. Basic Books, New York (2008)
11. Randall, M., Zirkle, C.: Information Technology Student-Based Certification in Formal Education Settings: Who Benefits and What is Needed. Journal of Information Technology Education 50(4), 287–306 (2007)
12. Scheuermann, F., Pedró, F. (eds.): Assessing the Effects of ICT in Education: Indicators, Criteria and Benchmarks for International Comparisons. European Union/OEDC (2009)
13. Stone, J., Madigan, E.: Inconsistencies and disconnects. Communications of the ACM 50(4), 76–79 (2007)
14. Wilson, C., Sudol, L., Stephenson, C., Stehlik, M.: Running on empty: the failure to teach K/12 Computer Science in the digital age. Tech. rep. ACM, New York (2010), http://www.acm.org/runningonempty/

A Competence-Oriented Approach to Basic Informatics Education in Austria

Peter Micheuz

Alpen-Adria-Universität Klagenfurt,
9020 Klagenfurt, Austria
peter.micheuz@uni-klu.ac.at

Abstract. Up to now, formal education in digital technologies at the lower secondary level of Austrian schools did not keep pace sufficiently with the requirements of our computerized society. But there are indicators to improve the unclear situation for pupils aged between 10 and 14 years. A reference framework for digital competence, embracing media education and basic Informatics education as well, is currently developed. After a selected look at various existing approaches, this paper deals with an Austrian project in that field and its current state.

Keywords: Reference Framework, Informatics Education, Digital Competence, Curriculum, Lower Secondary Education.

1 Introduction

Since the late 1980s, Informatics, ICT and Digital Media education at the lower secondary levels (grades 5 to 8) in Austria have shown a very inconsistent picture. Despite the outward appearance of a "digital patchwork", this important segment of the Austrian school system represents still a lively, although too fragmented scene. Many ambitious local and regional Austrian initiatives cannot hide the fact that our highly digitally penetrated society at the beginning of the 21st century is still not adequately represented in each Austrian obligatory school. One of the main reasons might be the fuzzy terminology in the field and its confusion with digital cultural techniques and digital media technologies. Another is the lack of a national curriculum and that of educational standards for this age-group. Finally, we can observe an inherent slow process of innovations of educational systems in general.

This paper outlines the status quo and current efforts relating to the development of a sustainable and coherent framework, including media education, ICT, and Informatics for all Austrian pupils at the lower secondary level. It follows on the author's contribution for ISSEP 2006 [1] which discarded the dialectic process between autonomy on the school level and the upcoming issue of national educational standards. Besides some major initiatives, e.g. the UNESCO/IFIP curriculum [7],[8], the ACM K-12 model curriculum [2],[9], and the framework "Principles and Standards for Informatics", published by the German Society for Informatics GI [3],[10], there are many regional initiatives to legitimize and structure this major educational concern. For historical and political reasons, Switzerland's 26 cantons

I. Kalaš and R.T. Mittermeir (Eds.): ISSEP 2011, LNCS 7013, pp. 43–55, 2011.
© Springer-Verlag Berlin Heidelberg 2011

and Germany's 16 federal states put much energy in establishing ICT standards separately. The Austrian situation at the lower secondary level is currently even more fragmented. Due to the lack of a national ICT framework and curriculum, schools and teachers act independently, teaching – if at the lower secondary level at all – Informatics and ICT according to school specific curricula. As an undesired consequence, schools and pupils proceed and perform at extremely different paces.

This paper describes a project of an Austrian ministerial task force to improve and consolidate this unsatisfying situation.

2   The Austrian Case Study

The Austrian school system encompasses elementary (grades 1-4), lower secondary (grades 5-8), and upper secondary level (grades 9-12/13). At the lower secondary levels the Austrian school system is divided into two types of obligatory schools, namely secondary general school (Hauptschule, HS) and the secondary academic school (Gymnasium, Allgemeinbildende Höhere Schule, AHS). Since two years, there is a large pilot project, called new middle school (Neue Mittelschule, NMS), exploring (without "out") new pedagogical approaches. According to current political intentions, all HS will be converted into NMS and the traditional Gymnasium will remain as a school type in its own right also in the foreseeable future. Currently about two thirds of the pupils attend the HS (NMS) and about one third attend the lower level of the AHS for four years. The reference model presented in this paper refers to all pupils aged 14 years at the end of the lower secondary level.

2.1   Anamnesis and Diagnosis

Since the late 1980ies, the elective subject Informatics has been implemented in the 7th and 8th grade of many lower secondary schools for interested pupils, dependent on engaged teachers. Considering the circumstances and restrictions of that time, the following development and proliferation of Informatics/ICT education can be regarded as successful. It was the time of pioneers among teachers and pupils when the "magic of the beginning" took effect and the upcoming curricular autonomy made possible an increased offer of Informatics/ICT classes. Within the process of profile building, Informatics and ICT have proliferated, and some schools still profit from this spirit of optimism until today.

Moreover, since 1990 an integrative use of ICT has been enacted by the Ministry of Education especially for the subjects German, English, Mathematics and Geometric Drawing, the so-called "Trägerfächer". This approach of implementing ICT at the lower secondary level for about ten years had good intentions, but it was not really effective. Interestingly, at that time the debate of Informatics as a subject in its own right and the integration of ICT in other subjects caused less confusion than today.

However, at the turn of the millennium the Austrian Ministry of Education has missed to anchor (basic) Informatics education at the lower secondary level within a new curriculum. Within the curriculum 2000, which is still valid, the only curricular reference for the overall use of ICT is expressed in a few sentences: "*Innovative information and communication technologies and mass media penetrate increasingly areas of life. Particularly multimedia and telecommunication are determining factors for the evolving information society. As part of teaching, these developments have to be taken into account, [...] and the educational potential of ICT has to be harnessed."* This short but clear message (*"has to be"*) was to ensure that all Austrian pupils at the lower secondary level should be provided with a minimum of ICT education in an integrative way in many subjects. The curricula for an independent elective or obligatory subject within curricular school autonomy remained very vague as well.

Now it is the year 2011 and – not surprisingly – due to this de facto non-existent curricular guidelines, the secondary level presents itself as a very inhomogeneous field with respect to Informatics/ICT. The subject Informatics/ICT is offered in different age-groups from grade 5 to 8 with different denotations, forms and amounts of hours per week. Additionally, this variety is expressed in different synonyms and denotations for the same subject. Within this scope one can find Informatics as well as information technology, introduction to Informatics, or even word processing and keyboarding. Still about one third of the pupils leave the lower secondary level without any formal education in this field as many schools still fail to ensure digital literacy for all in this age-group. Due to little efforts of Austrian's educational policy to reduce the digital divide between schools, the integrative approach of ICT penetration at the lower secondary level cannot be regarded as successful either. This patchwork with inconsistencies, disparities and digital gaps among pupils, schools and regions is well underpinned by empirical studies [1],[4],[18].

2.2 Therapy and Vision

From a "medical point of view", the Austrian patient with some insufficiencies with respect to a successful implementation of Informatics and ICT at the lower secondary level apparently needs some therapy. Metaphorically speaking, the patient needs a balanced "nutrition plan" in order to avoid being nurtured by one-sided food or even starving. Referring to basic Informatics education at the lower secondary level, this means a comprehensive and holistic set of content areas and concrete goals, based on the educational vision that all pupils at the end of lower secondary education should have acquired a certain level of digital competences which are discussed in this paper.

It seems like a good time to finally develop a long-awaited consistent national framework for digital competence based on Informatics education in Austrian's lower secondary education. The reason is that in European educational systems, especially in the German speaking countries Germany, Switzerland and Austria and not least triggered by the PISA studies, a remarkable debate and shift paradigm in the context of educational governance took place. In terms of curriculum development processes, PISA has led to a growing importance of educational standards, outcome control, competence orientation, and external assessment. Moreover it can be observed that PISA has contributed to a rediscovery of comparative education research [25]. There are EU-wide intentions to reinforce comparative studies with respect to Digital Competences which are, for instance, demanded in the European Reference Framework for Key Competences for Lifelong Learning [15].

After more than twenty years, the vision is to finally come to a national agreement on clear and binding objectives within a sound and acceptable framework of

Informatics/ICT related content. The time is now ripe for such a framework; during the last decade one could observe a boost in the digital penetration of our society which exerts much pressure on the education system. There is no question about clearly defining the role of general education in reacting to this challenge. What needs to be done, in medical terms, is to find an effective form of therapy based on clear expectations. The current Austrian approach and strategy is explained in the next chapters.

3 Clarifying a Fuzzy Field of Terminology

In contrast to traditional subjects such as German, English, Mathematics and Natural Sciences, all with elaborated curricula for each age-group, this is not the case at all for the comparatively new area of Informatics, ICT, and new media. Additionally, there is the inherent problem of a fuzzy terminology and contexts of meaning in the realm of digital technologies and their underlying educational values.

3.1 Informatics Education, Digital Competence or Media Literacy?

Since its emergence in the early 1990ies and long before the digital revolution has begun, the term "Informatics Education" has comprised three mutually interdependent views on computers as

- objects of teaching and learning activities (Informatics),
- tools for processing digital information (ICT as a cultural technique), and
- media for teaching and learning (E-Learning, Digital School).

Hubwieser [20] emphasizes the importance of an integrated view of computers in education. Basic Informatics education ("Informatische Grundbildung"), therefore, can be seen as an integrative amalgam of learning about informatics concepts, acquiring basic ICT-skills and using computers as networked multimedia devices responsively and effectively as a teaching and learning aid.

Thinking of a holistic view of Informatics education – as it expands to include more multi-disciplinary facets –, it is important to embrace an outward-looking view in computing that sees basic Informatics education as a field actively seeking to work with and integrate into other disciplines and areas. Such a view suggests the definition of frameworks as a means for promoting multi-disciplinary approaches while maintaining a clear identity of the field [11].

Based on the European Reference Framework for Key Competences for Lifelong Learning [15], where digital competence ranks after the first and foreign language, Mathematics, and basic competence in science and technology in fourth place, the EU Commission published the so called Digital Agenda for Europe [16]. According to this agenda, these key competences are indispensable to contribute to a successful life in the knowledge society. Competence in the fundamental basic skills language, literacy, numeracy, and in information and communication technologies (ICT) is seen as an essential foundation for all learning activities. A number of competences are not enumerated explicitly, but critical thinking, creativity, problem-solving, and risk assessment play a role across all eight key competences.

More specifically, the Digital Agenda defines digital competence *"as the individual capability to use Information Society Technology (IST) confidently and critically for work, leisure and communication, underpinned by basic ICT-skills which comprise retrieving, assessing, storing, producing, presenting information and participating in collaborative networks via the Internet."*

Consequently, the importance of IT and media literacy for students of all ages is evident and a clear message for education systems. *"It is essential to educate European citizens to use ICT and digital media and particularly to attract youngsters to ICT education. The supply of ICT practitioner and e-business skills, i.e. the digital skills necessary for innovation and growth, needs to be increased and upgraded. This calls for multi-stakeholder partnerships, training systems."* Accordingly, the Austrian Ministry of Education recently revised its ICT strategy "Digital Literacy in Austrian Schools" and launched the program "efit 21 - Digital Agenda for Education, Arts and Culture" [26]. But, as it is the case with many agendas, it lacks any concrete useful hints for a framework of digital competence at the lower secondary level.

3.2 More of the Same? – Living with Plastic Words

Currently, and this makes the allocation of ICT in the educational context so difficult, we realize an inflation of an arbitrarily exchangeable terminology, ranging from digital skills to informatics education wherein Informatics serves for almost every activity with computers at schools. In order to point out the problem of an inflationary and thus confusing terminology, Table 1 can serve as a phrase monger. The monuments of neologisms which can be created by combining these terms are often seemingly more important than the discussed objects themselves. At the same time, however, this confusing verbosity in form of so called "plastic words" [5] with unclear definitions offers the opportunity to create and form something new.

Table 1. Template for a phrase machine in a digital educational context

Prefix	Levels
Digital, Media Computer, E (Electronic) IT, ICT, **Informatics**	**Skills,** Literacy, Fitness Fluency, Knowledge, Qualification **Competence,** Pedagogy **Education**

Beside this diversity of notions, Informatics is still in a process of defining itself. This phenomenon is not only restricted to some European and Asian countries where the term "Informatics" is widely used, but also to "computer science". Tucker [6] considers severe public misperceptions about CS due to its confusion with programming, computer literacy and information technology. When developing frameworks and curricula, being aware of and coping with this complex and confusing terminology is necessary.

3.3 Competence Models in Context

Modern education systems and new curricula are based on competences, which means that the students' applicability of knowledge and skills is emphasised. The acquisition of competence and the degree of achievement is measured by completing tasks and solving problems. According to Weinert [12], competences include skills, knowledge and motivation to cope with new situations. Nowadays, the buzzword competence is accompanied by a vast plethora of publications. "It is not sufficient to know, you also have to apply it. It is not sufficient to will, you have also to do it." It was the famous German poet J.W. Goethe who already expressed this, about 200 years ago.

Competence models cannot be seen in isolation. They play a well-defined role on the long way from abstract objectives to effective classroom and learning activities resulting in verifiable learning outcomes. Normally, they are deduced from a core curriculum and form the basis for "educational standards". Such standards have been developed for other (main and PISA-relevant) subjects such as Mathematics. They are based on a particular process model respectively procedure [27].

1. Curriculum/Core Curriculum
2. Competence Model
3. Educational Standards
4. Tasks for Evaluation of Efficacy
5. Lesson Design and Teaching Methods

In addition to the problems of a fuzzy terminology, the introduction and embedding of ICT in lower secondary education by educational standards is formally not feasible for missing central curricula. Some regionally developed input-oriented curricula cannot hide the lack of a common nation-wide understanding about teaching objectives in that field.

In contrast, a competency-oriented approach aims at defining concrete learning outcomes for the end of secondary education. These competences to be acquired by the pupils have to be illustrated by tasks and should be principally assessed by methods of testing. And that's exactly what educational standards are about: clear educational goals, competences and tasks, and evaluating and checking by tests.

Unlike the nebulous situation at lower secondary education, educational standards are already well advanced in Austrian vocational schools (Berufsbildende Höhere Schulen, BHS). Constituting an important part of Austrian's upper secondary level, they offer their students a considerable amount of ICT and applied Informatics lessons nearly for all age-groups within elaborated and adapted curricula. Based on these curricula all steps of the process of educational standardization of outcomes have been taken [22]. Thus, these educational standards left already the status of theoretical concepts. Currently, in one of these school types one can observe already initiatives to evaluate their efficacy in school practice.

4 Structuring Objectives and Contents in the Field

A look beyond the national borders reveals worldwide efforts to propagate and structure the field of computer science and Informatics education in form of educational frameworks, curricula and standards. The complexity and diversity in this field is impressive and can hardly be overviewed in its whole bandwidth and depth. Long before the terms competence and educational standards dominated the educational debate in many European countries, the US report "Being Fluent with Information Technology" pointed already in this direction. This framework is in full compliance with the current European shift to a competence-oriented view of educational outcomes. Beyond mere skills training, competent students should acquire a deeper level of conceptual understanding that allows them to apply their knowledge of information technology to solving new problems in new domains and to learn to use new software as it becomes available. According to the already more than ten years old FITness program [13], competence is founded on three pillars:

- *Contemporary skills*, the ability to use various computer applications;
- *Foundational concepts*, the basic principles and concepts of computing that form the basis of computer science; and
- *Intellectual capabilities*, the ability to apply information technology in particular situations and to use this technology to solve new problems.

Contemporary skills change over time with the advances of software, while the underlying concepts remain stable. Intellectual capabilities are not restricted to single courses but should be developed throughout the curriculum in a cumulative way. Another seminal US initiative to improve the situation at the K12 level, the ACM curriculum 2003, yielded a pragmatic and readable taxonomy for the age-group K8 which complies with the lower secondary level in Austria. As we will see later on, this classification scheme should exert some influence on the Austrian approach.

- *Computers and software applications*
  [as knowledge of the computing environment]
    Parts of a personal computer, standard software, Operating systems, Networks, World Wide Web and E-Mail
- *Problem solving with computer science*
  [as a way of thinking that uses computers as a creative medium for solving problems of all kinds]
    Representing information digitally, Problem solving and algorithms, Computer programming
- *Social context of computing*
  [as an appreciation for the complex and changing interactions between computing, individuals, organizations, and culture]
    Privacy and security, evaluating and using information from networked sources, human-computer interaction, computers in society.

5 The Austrian Approach

5.1 Preliminary Considerations

Lower secondary education must be regarded as a window of opportunity and important phase of formal basic informatics education. Standardized learning objectives with clear expectations for teachers and students, based on a consistent, coherent, and outcome-oriented reference framework, are overdue.

In contrast to the German standards model [14] as a result of an informal group of informatics experts and teachers, an Austrian task force is supported officially by the Ministry of Education. Therefore, provided that the result in form of a framework with clear visions and expectations will be widely accepted by all stakeholders, especially by teachers, a successful dissemination into daily school practice should be realistic.

It is no secret that the Digital Agenda published 2010 by the EU Commission [16] played a substantial role to put the task force in charge, consisting of representatives of informatics didactics and school boards, and teachers as well.

Herein, digital competence requires a *"sound understanding and knowledge of the nature, role and opportunities of ICT in everyday contexts. This includes main computer applications such as word processing, spreadsheets, databases, information storage and management, and an understanding of the opportunities and potential risks of the Internet and communication via electronic media. Individuals should also understand how ICT can support creativity and innovation, and be aware of issues around the validity and reliability of information available and of the legal and ethical principles involved in the interactive use of ICT."*

With the same concern with which media educators [17] object to a prevalent technical view on digital media, computer scientists fear a dilution of their discipline. What both sides have in common is their complaint about the under-representation of their disciplines in formal education. Reiter elaborates in [19] extensively on the interdependency between digital media and informatics education and, conclusively, considers a combination of both innovative areas as a historic challenge. He envisages even the birth of a new obligatory subject. "Basic Informatics Education" or "Informatics and Digital Media" could be options.

The task force had to meet all demands and aimed to develop a balanced set of contents and objectives for all target groups as policy makers, school administrators, teachers, parents, and pupils who should be the main beneficiaries.

5.2 The Classification Scheme

After two years of occasional meetings and reviewing regional, national and international curricula and frameworks, the Austrian task force decided to develop a new balanced competence model and framework as a sound compromise of informatics and media education. It can be considered to be equivalent with the Austrian concept of „educational standards" for traditional subjects at the lower secondary level. Moreover, it can be seen as a supplementary and necessary part of the „Grundbildungskonzept" (concept of basic education) [21] in full compliance with the EU definition of key competences. Finally, it can serve as a solid fundament and preliminary stage for further Informatics and ICT teaching at the upper secondary level. After many years during which an overall concept has been lacking at the lower secondary level, it makes sense to build the house systematically from the first floor.

Table 2. Classification Scheme for Austria's Lower Secondary Level (K8 – 14 years) Competence Matrix for Basic Informatics Education

			Competence Levels		
			Basic	Extended	Special
Media Reflexion Related Topics	**1.**	**Information Technology, Human and Society**			
	1.1.	Impact of IT in Society			
	1.2.	Responsibility in Using IT			
	1.3.	Privacy and Data Security			
	1.4.	Developments and Vocational Perspectives			
Digital Media Knowledge	**2.**	**Informatics Systems**			
	2.1.	Technical Components and their Use			
	2.2.	Design and Use of Personal Information Systems			
	2.3.	Data Exchange in Networks			
	2.4.	Human-Machine Interface			
Use and Production of Digital Media	**3.**	**Software Applications**			
	3.1.	Documentation, Publication und Presentation			
	3.2.	Calculation and Visualization			
	3.3.	Search, Selection and Organisation of Information			
	3.4.	Communication and Cooperation			
Principles and Computational Thinking	**4.**	**Informatics Concepts**			
	4.1.	Representation of Information			
	4.2.	Structuring of Data			
	4.3.	Automatization of Instructions			
	4.4.	Coordination and Controlling of Processes			

This classification scheme is a German-Swiss-Austrian co-production of a working group, coordinated by Steffen Friedrich and the author at a seminar in Königstein (Germany, March 2011), and can be regarded as an amalgam of multi-perspective reflections and the incorporation of many aspects. It is integrative and consistent as well as interdisciplinary and multidisciplinary in its orientation. The working title is "Digital Competence" - "Basic Informatics Education", with the latter being more general. All empty cells are assigned with descriptors as concrete learning objectives in form of pupils-centered "I can …" statements. The descriptors of the basic level are currently approved by the task force.

At first sight, a structural similarity with the prominent Common European Framework of Reference for Languages (CEFR) can be observed. A closer look reveals also the influence of Baacke's seminal reflections on media competency [23]. The three levels of competence (basic, extended, special) refer, but are not equal in detail, to the Bloom's Taxonomy of learning objectives [24]. They describe the continuum from lower to higher order thinking skills. The tasks which have to be developed now have to take these different levels of difficulty and complexities into account. Finally, it is important to note that the consecutive numbers 1 to 4 of the content areas do not necessarily correlate with their importance.

5.3 Discussion of the Structure and Exemplary Descriptors

Information technology, Human and Society. Highly recognized and demanded by media educators, sometimes depreciated as "soft Informatics" by technically oriented teachers, this content area has become increasingly important during the last decade. Every informed citizen has to find his digital identity, to reflect on his responsibility and to evaluate chances and risks in the information society. European initiatives such as Klicksave in Germany and SaferInternet in Austria, aspects of vocational education, and also historical subject matters have their place here. To give an impression of the model's granularity, some exemplary descriptors in form of "I can" –statements are:

- I can enumerate areas where the computer cannot replace a human.
- I can distinguish between data protection and data security.
- I can evaluate the effects of my behavior when playing online games.

Informatics Systems. In order to be a competent user, a profound media knowledge of networked hard- and software is indispensable. This applies not only to the decreasing prevalence of traditional PCs, but increasingly to mobile devices and the upcoming cloud computing. The design of the framework is future-proof as the descriptors do not draw on specific hardware and operating systems.

Examples for the pupils-centered statements are:
- I can use common input devices as the keyboard quickly.
- I can enumerate different storage media.
- I can explain the difference between the Internet and the WWW.

The issue "fluent keyboarding" is often a matter of debate, especially among some informatics teachers who do not attach much importance to it. Due to this framework, pupils are expected to acquire a reasonable degree of keyboard handling in formal education. In contrast to the explanation of the difference between Internet and WWW – doubtlessly an intellectual challenge with a historic background –, keyboard fluency is the result of manual training and, as a practical skill, can be very easily tested.

Software Applications. This content area, together with Informatics Systems, must be considered as the core of the framework. It represents the new cultural techniques which are already covered in many current ICT lessons. Fluent digital media use and production imply e-skills which can be acquired only through regular training in various contexts. This area is widely overlapping with the syllabus of the application- and product oriented European Computer Driving License, a certificate which plays a considerable role at the Austrian lower secondary level. The framework goes beyond the low level skills of the ECDL Core, including a competency-oriented approach. Examples for objectives in this wide area of skills and competences are:

- I can design digital documents including texts and pictures.
- I can describe the basic structure of a spreadsheet.
- I can use E-Mails and forums for exchange of information and cooperation.

Informatics Concepts. In this content area (real?) informatics teachers and informatics experts feel comfortable and Informatics in its core - not to be confused with Basic Informatics Education - as the referring discipline of information technology is prevalent. Here, concepts like digitalization, algorithmics and even programming on an age-appropriate level come into play. All of them aim at a deeper understanding of the field. We have to be aware that not all pupils are able to fulfill the demands of this cognitively challenging area. For pragmatic reasons, an abstract approach and formalism, as observed in other countries such as Bavaria has been avoided here.

Examples of operationalized objectives are:

- I can explain the input-processing-output principle.
- I can enumerate important data types.
- I can create simple programs in an appropriate development environment.

Additionally, regional initiatives such as "Experiencing Informatics", the Austrian adaptation of CS Unplugged, robot contests, and the Beaver Contest are good examples to broaden the picture of informatics education as an intellectual challenge beyond the consolidation of computer literacy as a cultural technique.

5.4 Further Reflections and Actions

Whilst the classification scheme in Table 2 with four main categories and four content areas each, together with about seventy "I can ..." descriptors, is approved for the basic level as the minimum standard for all pupils aged 14 years, the formulation for the extended and special competence levels have yet to be finalized. Prototypical tasks and assignments to illustrate and concretize the competences have to be developed, published, tested and calibrated as well.

Curricula can be regarded as results of cultural traditions and findings from science and empirical research, and not least from framework conditions given by educational policy. The competence framework presented in this paper, and published also on the web at **http://www.informatische-grundbildung.com**, has been developed without referring to a valid national curriculum. As a consequence and for the time being, it has to be considered still informal and with no obligations for schools, teachers and pupils. However, the orientations implied by this concept are promising.

One function of this model is to provide schools with guidance in the implementation of binding educational objectives. These can serve as a road map for policy makers, teachers, pupils, and parents as well. An other function is to make it possible to assess and evaluate educational outcomes in context of basic Informatics education and thus to determine whether the pupils have acquired the desired objectives. The competence matrix can also provide orientation for individual diagnosis and supplementary support measures.

Provided, that there will be a broad agreement on the reference model and its standardized learning objectives, the autonomous schools then will be faced with the challenge of effectively and efficiently meeting them. Integrating ICT in existing subjects and/or implementing a new (interdisciplinary) subject will be a key question. Another task will be the supply with enough competent teachers and the development of competence-oriented curricula, and teaching material for the grades 5 to 8.

6 Conclusions

The call for improving the situation of a fragmented and confusing, although occasionally and regionally blossoming, Austrian landscape of basic Informatics education at the lower secondary level has been growing increasingly louder. As a consequence, the Austrian Ministry of Education charged a task force to develop a reference model balancing the demands in a very complex field ranging from digital media to basic Informatics education. Being still work in progress, these efforts of consolidation will need the acceptance of all stakeholders. Still a lot of work in form of finalizing the extended/expert competence levels, developing tasks, evaluating and testing has to be done. Finally, a strategy for taking effect in schools has to be found.

At the moment, this preliminary reference framework has to be seen as a starting point for a nationwide discussion among policy makers, teachers, parents, and pupils. Finally, when the framework is revised and ready for preliminarily dissemination, it should serve as a guide for schools, teachers, and parents to be clear about the degree of computer literacy, digital competence and/or informatics education (cf. chapter 3.2) which pupils are expected to have acquired at the end of lower secondary level.

The results of the task force should not be only a valuable tool in the hands of academics. Otherwise this work is nothing but an inert mass of words and phrases, as one of the members of the task force put it: *"I believe, it would be better to be modest in the minimum requirements - so that the framework appeals to teachers and pupils. This is better than to satisfy all unrealistic requirements from experts. Teachers could then easily turn away before they really deal with the particular objectives of the framework. The worm has to taste good to the fish, not to the fishermen. Everybody can do more in class, but as many (teachers) as possible should do the little which seems to us, the experts, indispensable."*

Last but not least, this reference model shall impart a vision of digital education processes and incorporate a modern and clear philosophy of the fuzzy digital area at the lower secondary level. It can offer prospects for the development of contemporary and future-oriented competences and has the potential to become soon a driving force in the digital development of our lower secondary schools.

An old proverb says: "The proof of the pudding is in the eating". The pudding is currently cooked and will presumably be served soon. Hopefully, it will be eaten by all teachers and pupils concerned.

References

1. Micheuz, P.: Informatics Education at Austria's Lower Secondary Schools between Autonomy and Standards. In: Mittermeir, R.T. (ed.) ISSEP 2006. LNCS, vol. 4226, pp. 189–198. Springer, Heidelberg (2006)
2. Association for Computing Machinery (ACM), http://www.acm.org (March 31, 2011)
3. Gesellschaft für Informatik (GI), http://www.gi.de (March 31, 2011)
4. Micheuz, P.: The Role of ICT and Informatics in Austrian Secondary Academic Schools. In: Mittermeir, R.T. (ed.) ISSEP 2005. LNCS, vol. 3422, pp. 166–177. Springer, Heidelberg (2005)
5. Pörkson, U.: Plastikwörter - Die Sprache der internationalen Diktatur. Klett-Cotta, Stuttgart (1989)
6. Tucker, A.B.: K-12 computer science: aspirations, realities and challenges. In: Hromkovič, J., Královič, R., Vahrenhold, J. (eds.) ISSEP 2010. LNCS, vol. 5941, pp. 22–34. Springer, Heidelberg (2010)

7. Anderson, J., Weert, T.: Information and Communication Technology in Education. A Curriculum for Schools and Programme of Teacher Development. Division of Higher Education, UNESCO (2002)
8. Unesco/IFIP Curriculum – ICT in Secondary Education (1994), http://wwwedu.ge.ch/cptic/prospective/projets/unesco/en/welcome.html (March 31, 2011)
9. A Model Curriculum for K–12 Computer Science: Final Report of the ACM K–12, http://www.acm.org/education/education/curric_vols/k12final1022.pdf (March 31, 2011)
10. Puhlmann, H., et al.: Grundsätze und Standards für die Informatik, http://www.informatikstandards.de (March 31, 2011)
11. Sahami, M.: Setting the Stage for Computing Curricula 2013: Computer Science Report from the ACM/IEEE-CS Joint Task ForceSIGCSE 2011, Dallas, Texas, USA (2011)
12. Weinert, F.E.: Vergleichende Leistungsmessung in Schulen – eine umstrittene Selbstverständlichkeit. In: Weinert, F.E. (Hrsg.) Leistungsmessungen in Schulen, pp. 17–31. Weinheim und Basel (2001)
13. Committee on Information Technology Literacy. Being fluent with information technology. National Academy of Sciences, Washington, DC (1999)
14. Gesellschaft für Informatik (GI) e.V.: Grundsätze und Standards für die Informatik in der Schule, Bildungsstandards Informatik für die Sekundarstufe I. Addendum to LOG IN 28, 150/151, Berlin (2008)
15. European Reference Framework for Key Competences for Lifelong Learning European Communities, Belgium (2007), http://ec.europa.eu/education/index_en.html (March 31, 2011)
16. Digital Agenda, http://ec.europa.eu/information_society/digital-agenda/documents/digital-agendacommunication-de.pdf (March 31, 2011)
17. Medienpädagogisches Manifest, p. 2, http://www.keine-bildung-ohne-medien.de/medienpaedagogisches-manifest.pdf (March 31, 2011)
18. CDA Sonderhefte, http://www.box.net/sonderhefte (March 31, 2011)
19. Reiter, A.: Medienbildung auf Überholspur. In: Brandhofer, G., et al. (eds.) 25 Jahre Schulinforma-tik in Österreich, Wien. Österreichische Computergesellschaft, Band, vol. 271 (2010)
20. Hubwieser, P.: Didaktik der Informatik. Springer, Berlin (2003)
21. Grundbildungskonzept, http://fbm.uni-klu.ac.at/lgmodule/ModulNAWI.pdf (March 31, 2011)
22. Bildungsstandards für Berufsbildende Schulen, http://www.bildungsstandards.berufsbildendeschulen.at (March 31, 2011)
23. Baacke, D.: Medienkompetenz als zentrales Operationsfeld von Projekten. In: Handbuch Medien, Bonn, pp. 31–35 (1999)
24. Bloom, B.S. (ed.): Taxonomy of Educational Objectives, the classification of educational goals – Handbook I: Cognitive Domain. McKay, New York (1956)
25. Ertl, H.: Educational Standards and the Changing Discourse on Education: The Reception and Consequences of the PISA Study in Germany. Oxford Review of Education 32(5), 619–634 (2006); Special Issue: Comparative Inquiry and Educational Policy Making
26. BMUKK, eFit21-Strategie, http://www.elearningcluster.com/pdf_s/erlass_digitale_kompetenz.pdf (March 31, 2011)
27. Oelkers, J., et al.: Qualität entwickeln – Standards sichern – mit Differenz umgehen (2008), http://www.bmbf.de/pub/bildungsforschung_band_siebenundzwanzig.pdf (March 31, 2011)

Outreach to Prospective Informatics Students

Maciej M. Sysło[1,2,3]

[1] Faculty of Mathematics and Informatics, Nicolaus Copernicus University,
Chopin str. 12/18, 87-100 Toruń, Poland
syslo@mat.uni.torun.pl
[2] Institute of Computer Science, University of Wrocław,
F. Joliot-Curie str. 15, 50-383 Wrocław, Poland
syslo@ii.uni.wroc.pl
http://mmsyslo.pl/
[3] Warsaw School of Computer Science
Lewartowskiego 17, 00-169 Warsaw, Poland

Abstract. In this paper we first identify the main factors which influence the students' attitudes to study computer science related disciplines. Then various outreach initiatives and activities implemented in Poland are described and discussed. They range from changes in the national curriculum for middle and high schools to formal and informal lectures, courses, and workshops organized by public and private institutions of tertiary education. Project Informatics +, addressed to 15 000 students from five regions, is presented in details and its outcomes after the first year of running are reviewed shortly.

1 Introduction

In this paper, the term 'informatics' (pl. *informatyka*) is equivalent to 'computer science' and we use 'computer science' when we want to emphasize scientific aspects of the discussion. The term 'computing', however, which embraces 'computer science', 'software engineering', 'information systems', 'information technology' and some other computer-related titles, has no official counterpart in Poland. Recently, the term 'computational thinking' became popular.

This paper is a continuation of our works on informatics education in Poland presented at the previous ISSEP meeting.

In the paper [17] presented in Klagenfurt (2005) we focused on the question: how much informatics is needed to use information technology and how to prepare teachers for their new role as moderators of students' learning, to use information technology in various situations.

In the paper [18] presented in Toruń (2008), motivated by a gradual decline in the number of students applying to earn a computer science degree, we described a learning and teaching framework for schools which is aimed at increasing student interests in studying computer science as a discipline, or at least in better understanding how the computer and its tools work and can be used in solving problems which may occur in different areas. We are convinced that the learning methodology about computer use by students and applying computers and information technology to solving

I. Kalaš and R.T. Mittermeir (Eds.): ISSEP 2011, LNCS 7013, pp. 56–70, 2011.

problems is a good motivation and preparation for their future decisions to study computing and become computer specialists.

Both papers [17] and [18] refer to the situation in high schools in Poland in the mid 2010's, when information technology was a curriculum subject (1 hour per week for one year).

In this paper, with the same motivation as in [18], we present outreach[1] activities aimed at prospective students and also at the public about the importance of computer science knowledge and attractiveness of the computer related professions in the knowledge based economy and society.

In 2008 the national curriculum was modified and some of the changes were in favor of informatics education – we describe them in Section 3. It is expected that beginning of 2012 informatics education of all school students will be put on a higher level and students will be better prepared for considering computing as a discipline of their future study and professional career.

The society needs a continuous inflow of good students in computer science, science, and engineering to be educated and trained as professional specialists for informatics related jobs in order to sustain the developments and achievements that are necessary to meet the expectations of the information society and its citizens. Yet, the number of students opting for computer science education appears to be declining and various factors affect students' interests in these fields.

The methodology used in our activities with students is based on the idea of **computational thinking** (see [19], [2]). We are convinced, repeating after [18], that computational thinking could be added to the traditional three Rs: (i.e. reading, writing and arithmetic) as an additional basic skill needed especially by high school students – they will be better prepared to choose a future career as either a computer scientist or a computer specialist.

The paper is organized as follows. In Section 2 we shortly discuss social understanding of computer science with relation to its popularity as a discipline of study and interest among schools. In Section 3 we briefly describe informatics education in Poland. Section 4 is concerned with the main topic of this paper – we present a number of outreach initiatives and activities undertaken in Poland and describe their role in computer science education in schools and in the society.

2 Computer Science Education in Crisis

There is a general opinion that school students are not prepared to make a decision about their future career and professional life related to computer science. Moreover they misunderstand what computer science really is. In [18] we have extensively discussed main factors which have caused for the last 3-5 years a substantial decrease in the number of computer science enrollments – for instance it was estimated that it dropped in half in the USA. Here, following [18], we briefly repeat some of the arguments which are important for our discussion about outreach activities.

---

[1] **Outreach** is an effort by an organization or a group (here curriculum teams, universities) to connect its ideas and practice to the efforts of other organization (here schools), groups, specific audiences, or the general public.

Many people, among them education policy makers, teachers, academics and parents, do not consider computer science as an independent science and, therefore, as a separate school subject. Most of them confuse computer science and information technology and limit informatics in education to provide students and teachers with computers and Internet access.

Informatics education in school does not clear up the myths about computer science, for instance it is still confused with computer programming. Students have access to high-level tools for designing and producing complex applications without any knowledge of fundamentals of computer science such as logic, discrete mathematics, programming methodology, or computability.

Today almost all students have computers and access to Internet at home. Therefore most of high school graduates are quite fluent in using computers to play, search the web and communicate and, as a result, they have no real interest in pursuing computer science as a career choice. They have tasted enough information technology while growing up and want something different at the university level. To change this, informatics classes should prepare students for further study instead of being satisfied with the knowledge and skills they have already learned.

Youth's infatuation with technology does not extend to their desire to learn the discipline of computer science – one of our goals in outreach activities is to motivate students to go 'beyond the screen' and investigate how computers and software work so they can create their own computer solutions.

One of the challenges to a curriculum in computer science is to catch up to the new technology and to adjust it to rapidly changing markets and users' expectations. There is no longer a need for a large number of computer scientists working on foundations of the discipline and developing basic products as it was in the 1960's and 1970's. However there is still a demand for experts and specialists in various areas of computer use and applications who are competent in the range of the university curriculum in computer science. A computing degree can help to find a job in science, engineering, health care, finance, and so on. The availability of jobs, as well as the impact of computing in society should motivate students to study computer science.

The White Paper by the CSTA [16] lists a number of challenges and requirements that must be met if we want to succeed in bridging the gaps in education and improve education in informatics as a computer science discipline:

- students should acquire a broad overview of the field of computer science;
- informatics instruction should focus on problem solving and algorithmic (computational) thinking;
- informatics should be taught independently of specific application software, programming languages, and environments;
- informatics should be taught using real-world problem situations;
- informatics education should provide a solid background for the professional use of computers in other disciplines.

One of our goals in this paper is to show how we partly meet these challenges in our approach to informatics education for all students in schools in Poland, and enhance and support schools by some outreach activities addressed to prospective computer science students.

3   Informatics Education in Poland – In the Past and Today

In the education system in Poland, **informatics education** consists of two types of classes and/or activities:

- separate informatics classes;
- across-curriculum integration of computers, information and communication technology, and Internet with learning and teaching of all subjects.

Detailed information about the development of informatics education in Poland is included in [18]. Here we refer only to the most important steps, in particular we report on changes in the curriculum introduced by the reform at the end of 2008.

The first informatics classes in Poland were organized in the mid sixties. The main topics of instruction were algorithms for numerical calculations and programming in Algol 60 – algorithmics was restricted to numerical methods. The first national curriculum for informatics as an independent subject was proposed in 1985. In the mid 90's, the term 'information technology – IT' (later 'information and communication technology – ICT) was accepted by the education policy makers in Poland and a new subject information technology was introduced to the curriculum by the Education Reform of 1997 and as a result information technology became the high school independent subject in 2002. Informatics as separate subject for all students has returned to high schools as a result of the reform of 2008.

It is interesting to note that informatics as a separate subject has been in the national curriculum and in the schools in Poland since its introduction in 1985. The author is not aware of any other such country.

3.1   Education System in Poland

For a long time formal education in Poland started at the age of 7, which has recently been lowered to 6. Since 1999 the school system at the primary and secondary levels has consisted of three stages:

- primary school – 0-6 grades (age 6 to 13);
- middle school (in Polish: *gimnazjum*) – 7-9 grades (age 13 to 16);
- high school – 10-12 grades (to 13 in certain vocational schools) – (age 16 to 19).

Independent informatics subjects have been in our national curriculum since 1985 and recently have been modified by the reform at the end of 2008.

3.2   New Curriculum of Informatics

In what follows we describe in more details the actual curriculum of separate informatics subjects approved at the end of 2008 and introduced to primary schools (1-3) and to middle schools in 2008 and as it will be introduced to primary schools (4-6) and to high schools in 2012. The changes made to the existing curriculum are important to any outreach activities taken by a tertiary education institution. From one side, it may be assumed that school students are familiar with the topics listed in the curriculum, and from the other side – outreach activities should enhance and extend the curriculum topics. It is very important to connect outreach activities to what is actually taught in schools – one can easily lose attention and interest of school students when the topics are far from what they know and from what they can follow. As a teacher in the project Informatyka + (see Section 4.4), prior to my lectures and workshops, I usually ask school teachers who accompany students from schools how advanced their students are in algorithmics and programming. Similarly do other members of our staff in this project.

3.2.1 Primary Schools

In the previous curriculum, in primary schools there was informatics in grades 4-6, at least 2 hours per week for one year (or one hour per week for two years). In the new curriculum the separate informatics subject, called now **computer activities**, has been substantially extended and now it runs through grades 1 to 6.

In grades 1-3, computer activities do not form a separate subject and are supposed to be fully integrated with other activities like reading, writing, calculating, drawing, playing etc. – in fact there are no separate subjects in grades 1-3, only activities.

In grades 4-6, one hour per week for three years is assigned to computer activities. At this stage it has been already advised to teachers and expected from students that these activities follow the general approach to problem solving with computers described shortly in Section 3.3.

3.2.2 Middle Schools

In middle schools, as it was before, **informatics** is at least 2 hours per week for one year or one hour per week for two years.

The curriculum of informatics for middle schools contains, as before, a section on algorithmics, algorithmic thinking and solving problems with computers. Although programming is not included in the curriculum, an introduction to Logo or to another programming language is a part of the instruction in some schools and students from those schools take first steps in programming.

Within algorithmics, students are expected, as outcomes, to be able to (this part of the curriculum has been modified):

- explain what an algorithm is,
- provide a formal description (specification) of a simple problem situation and propose an algorithm for its solution;
- use spreadsheets to solve simple algorithmic problems (e.g. the change problem);
- describe, how to find an element in an ordered or an unordered sequence of elements;
- use a simple sorting algorithm (e.g. by counting);
- perform (run) some algorithms on a computer – either writing a program in Logo or in another language, with the help of spreadsheets or running an education software.

As a novelty, activities of students within the framework of Web 2.0 have been introduced to the curriculum such as students taking part in web discussions and publish their information and opinions. It is assumed that these activities of students are moderated by teachers.

A new textbook for informatics in middle schools [8], incorporating changes introduced by the new curriculum, was published in 2009.

3.2.3 High Schools

In the new curriculum for high schools **information technology** disappears as an independent subject and informatics has been introduced in its place, at least 1 hour per week for one year. In consequence, beginning of 2012, there will be also **informatics for all students** in high schools, as it is in middle schools.

Informatics (understood as computer science) remains in high schools as an elective subject and is taught only in some schools. Students may also take an external final examination (*matura* in Polish) in informatics.

Informatics for all students

Again, as in middle schools, the main emphasis is put on problem solving with computers using the methodology described shortly in Section 3.3. Problems may come from various fields, in particular from school subjects, and students may use a variety of informatics tools for solving them.

Students are expected, as outcomes, to be able to:

- discuss and analyze various problem situations;
- develop and formulate specifications of various problem situations;
- design a solution of a problem by choosing a solution method and computer tools, such as a programming language, application or education software;
- run a solution on a computer and test and evaluate its properties such as complexity (efficiency) and correctness with regard to the specification;
- present a solution and discuss its applications to other problem situations.

Additionally to problem solving skills, all students in high schools are expected to publish in the web their own educational content and use e-learning to enhance and enrich their learning environment by including open content and courses.

A new textbook for informatics for all students in high schools [9] will appear in 2012. We propose a project based learning as a working method and computational thinking as an approach to problem solving.

Informatics – elective subject

No significant changes have been introduced to the new curriculum of informatics as an elective subject. A new textbook [10] will be published in 2012 in which computational thinking will be more substantially involved as a working methodology.

3.2.4 Comments on Changes in the Curriculum

The reasons behind the changes described so far and also expectations of real changes in students and teachers behaviour are as follows:

- it is assumed that integration of computers with students activities in grades 1-3 will result in a habit of using computers as tools supporting learning of various subjects and disciplines at next stages of education, formal, non-formal, and incidental in school and at home;
- computer activities in grades 4-6 are supposed to lay down solid knowledge and skills within the range of information and communication technology to be used at the next stages of education, formal, non-formal, and incidental;

- informatics in middle schools is supposed to introduce basic elements of informatics, as computer science, important for at least two reasons: as a starting point for informatics education of all students in high schools and as a pre-orientation for those students who might be interested in choosing a high school which offers a specialization in computer science topics, such as algorithmics, networks, data base, etc.;
- introduction of informatics for all students in high schools has at least two main missions:
  - o this subject gives a feeling and touch of informatics as computer science to all students; although most of them will continue education, choose career, and find jobs in other disciplines, more and more careers and jobs become IT professions [5] which require a solid background in computer science and its applications;
  - o it is a continuation of pre-orientation, started in middle schools, intended to prepare school students for their choices of future study, career and jobs in computing related disciplines and fields;
- web activities constitute another area of informatics education from which the new curriculum benefits much due to the increasing role of the Internet in all activities: scientific, practical, and personal – it is also expected that high school graduates will be prepared to actively use the Internet as an e-learning environment for their lifelong learning activities.

3.3 Computational Thinking

Our approach to informatics (computer science) education in schools, from the beginning in the mid 60's till the beginning of the new era, was to put the emphasis on algorithms and **algorithmic thinking** as the main components of computer science. We have used a methodology, called **algorithmic problem solving**, for the systematic development of a computer solution for a problem, which covers the entire process of designing and implementing solutions, from beginning to end. This methodology is aimed at generating good solutions, characterised by the three fundamental properties: readability (the solution is understandable to anyone who is familiar with the problem domain and computer tools used), correctness (the solution satisfies the problem specification), and efficiency (the solution doesn't waste computing resources, time and space). The algorithmic problem solving approach consists of six stages and is described in details in [18].

A much wider view on computing competencies has been proposed by Jeannette Wing in her paper on **computational thinking** [19] – it extends algorithmic thinking and fluency in working with information technology to competencies which are built "on the power and limits of computing processes, whether they are executed by a human or by a machine."

Today in our approach to informatics education (see [18] for details) we adopt computational thinking as the main learning and teaching methodology about computer use and applying computers and information technology in solving problems. This approach can help students to add computational thinking to the traditional three Rs: (i.e. reading, writing and arithmetic) as an additional basic skill needed by everyone. This approach is also used in our outreach activities described in this paper as the way to better prepare our school students for their future decisions to study informatics related disciplines and to encourage them to consider a future career in computing.

Recently (see [4] and [2]), the computational thinking approach has been adopted as a methodology that can be used across all disciplines to solve problems and improve understanding of the power and limitations of computing in the modern age.

4 Outreach Activities

In Section 2 we have identified some factors which influence the students' attitudes to study computing related disciplines. In response, various outreach initiatives and activities are taken.

In the US, the Computer Science Teacher Association (CSTA) collected data [3] which shows that computer science is on the decline in high school. In the recent report [14], ACM and CSTA call for federal, state, and local actions and together with other parties they have formed a coalition **Computing in the Core** (CinC) to address the need to build a K-12 computer science program in the US schools. In November 2010, President Barack Obama announced the launch of several nationwide programs to help motivate students to master in STEM (Science, Technology, Engineering, Mathematics) related subjects, see also [15]. The NSF also has announced in 2010 the **CS/10,000 Project** [12] and proposed a new high school computing curriculum which will be taught by 10,000 newly prepared teachers in 10,000 classrooms across the US. Most recently CSTA has published standards [4] which provide a three-level framework for K-12 computer science education. In particular, the standards in the course *Computer Science in the Modern World* reflect learning content that should be mastered by all students, similarly to the subject informatics for all students in middle and high schools in Poland.

In the next sections we describe the most successful initiatives implemented in Poland in the last 2-3 years. They range from changes in the national curriculum for middle and high schools to formal and informal lectures, courses, and workshops organized by public and private institutions of tertiary education.

4.1 New National Curriculum

Informatics education in the new national curriculum has been described in Section 3.2 where we also emphasise the importance of changes in the curriculum to better prepare students for general education as well as for their future choice of a next step in education and professional life. The outreach activities described in this section quite often refer to the new curriculum by enhancing and deepening students' curriculum achievements and extending them in the case of talented and gifted students.

4.2 Workshops for Students and Teachers

It has been observed that students in middle schools and in high schools during informatics classes spend less time on programming, which is usually more time consuming, than on designing algorithmic solutions. Moreover, they are not encouraged to program by teachers who usually have no sufficient practice in programming.

Regional Informatics Circle (pl. *Regionalne Koło Informatyczne – RKI*), supervised by a group of academic teachers from the Faculty of Mathematics and Informatics, Nicolaus Copernicus University, is a novel approach to increasing programming skills among secondary school students in the Kujawsko-Pomorskie Region. It is fully based on distance learning and individual work of students after regular school hours. In this project the achievements of students are monitored on-line through weekly programming contests, which are supported by a system for automatic correctness verification and timing of students codes. The standardized tests have been developed to monitor students' skills. C (CodeBlocks) and Java (NetBeans) are used as programming languages and environments and the educational platform OLAT is a communication medium in this project.

Almost 1000 students participated in the project in the school year 2009/2010: 776 students participated in Part I (Programming in C or in Java – basic level), 190 – in Part II (Algorithmics – basic level), and 30 in Part III (Algorithmics and programming – advanced level). Detailed analysis of the project´s outcomes will be published elsewhere.

The Department of Informatics and Information Technology Education in the Faculty of Mathematics and Informatics, Nicolaus Copernicus University offers also regular in-service seminars and workshops for computer science teachers from secondary and high schools. Some recent topics of the seminar are: new curriculum, computational thinking versus algorithmic thinking, recursion, text algorithms, teachers' preparation standards, educational platforms, network administration.

High schools in Poland do not offer AP computer science courses. However, some universities encourage school students to participate in university courses counted sometimes as tertiary courses. The Faculty of Mathematics and Informatics, Nicolaus Copernicus University offers informatics courses which are attended by students from GiLA in Toruń, a middle and high school ran by the University. In 2010 the GiLA took the first place in two out of three rankings of secondary schools in Poland.

4.3 Competitions and Olympiads in Informatics

Competitions are typical outreach activities, they are usually ran by parties external to schools. These educational events require knowledge and skills exceeding what is taught at schools. They engage and develop skills necessary in the future professional activities such as: constant self-development, self-discipline, hunger for knowledge, ability to work in a team. No competition, however, is the goal for itself.

Olympiads in Informatics

The achievements of young Poles, school and university students, in international programming competitions in the past 15 years are well known [6]. The Polish experiences are universal enough to be adopted also by other countries and could help to work with students talented in computer science.

The Olympiad conducts intensive educational activities: post-Olympiad materials are published and contain detailed analysis of task solutions; former Olympiad contestants run a portal for beginners in the field of programming and algorithmics [21], finalists of the Olympiad participate in summer camps combining recreation and education. However, talented students are first discovered by their school teachers. The Olympiad organizes also workshops for teachers, where they practice how to work with talented students and prepare them for computer science competitions.

The Olympiad in Informatics for middle schools has been also established in Poland, although students from middle schools may participate in the Olympiad for high schools, and some of them are very successful [13].

The Bebras (Beaver) contest

The idea of Bebras contest was born in Lithuania, by Prof. Valentina Dagiene, where the first contest was organized in 2004. In 2010, Bebras took place in 14 countries, with about 235 000 participants. Some other countries – Israel, Cyprus, Japan, Malta, Russia – are interested to join the contest [1].

The main aim of the Bebras contest is to promote interest in information and communication technologies as well as in informatics (computing) to all school students. Moreover the contest encourages students to use modern technologies in their learning activities more intensively and creatively.

Bebras, like Kangaroo in mathematics, is a one stage contest addressed to school students of all grades. Tasks are on information comprehension, logical and algorithmic thinking, games and puzzles, graphical representations of notions and objects, computer and software functions, etc. Tasks touch also various school subjects and topics.

In ongoing research we want to learn whether and how Bebras results may be used to judge about informatics education and about the development of computational thinking skills through the consecutive stages of school education.

4.4 Project Informatics +

Informatics + is one of the largest outreach projects in Poland, see [11]. It is run by the Warsaw School of Computer Science (WWSI) [20], a private university established in 2000, one of the few private schools which offer a master degree in computer science. The author of this article coordinates scientific and education activities in this project. The project is financed by EU Funds. It is addressed to high school students in five regions (states – *województwo*) in Central and Eastern Poland and it is expected that more than 1000 schools, 15.000 students, and 300 teachers will participate in this project in 2008-2012.

Project goals

The main goals of the project Informatics + are as follows:

- elaborate and implement innovative methods of teaching and learning key competencies in informatics and its applications;
- improve and extend off school students' activities in developing key competencies in informatics and its applications;

- extend students' interests about job market expectations and better preparation for their future choices of professional development;
- extend opportunity talented and gifted students have to improve their informatics interests and competences, in particular those students who are interested in taking part in numerous informatics competitions;
- improve students' school achievements (measured by school grades) in informatics and in other related subjects;
- provide schools with open education content in informatics and its applications;
- introduce school students to an academic type of instruction which differs significantly from school lessons;
- develop teachers' competences in working with students talented in informatics.

Project organization

In this project students may participate in:

- lectures (2 hours);
- lectures (2 hours) followed by computer workshops (3 hours);
- extensive workshops (24 hours);
- competitions, such as: the Informatics Olympiad, Beaver, "Our school in Internet", on 3D graphics, and web contest;
- summer computing camps in an attractive spa – a combination of vacation activities with plenary lectures and discussions, workshops, and on-line competitions.

Moreover, in-service courses are also offered for teachers to improve and develop their competencies in working with students talented, gifted, and particularly interested in advanced informatics topics. Lectures and computer workshops are delivered in WWSI, in one of the five Regional Centers of the Project, and in schools.

The courses are offered on two levels:

- basic – addressed to all students, supposed to extend the curriculum knowledge in various informatics topics;
- extended – these are mainly extensive workshops (24 hours), addressed to students interested mainly in deepening their informatics skills.

The Educational platform Fronter [7] is used in the project as a communication medium, as an element of cloud computing (the platform is hosted in Oslo). It contains all course materials (lecture notes, presentations, education software, programming codes, etc) prepared by teachers and students use the platform to save their works done during workshops. Then students may use all these materials anytime and anywhere when they return home or to school and want to continue their work in class. The platform is also used to build and run tests and to collect students' opinions about the course they attend.

Informatyka + contributes also to the Polish Open Computer Science Platform (*Polska Wszechnica Informatyczna*) which is a collection of more than 60 lectures delivered by well known specialists in various areas of informatics and its applications (WWSI got a prize for this project) http://www.pwi.edu.pl/.

Project topics

The project Informatyka + consists of five thematic modules (we list also titles of some courses within modules):

1. **Algorithmics and Programming:** Searching and sorting – the power of order, Simple computer calculations – can all be computed, Algorithmic techniques, Shortest paths and trees, Data structures and their use (advanced), Advanced algorithms, Matura (final examination) in informatics.
2. **Data Base:** Data base – fundamentals, SQL language (basic and advanced level), XML documents in data base, Technology ADO.Net, Data mining, T-SQL language.
3. **Graphics, Multimedia, Internet techniques:** Graphics editor – GIMP, Working with multimedia, Searching for multimedia in the Internet, Creating dynamic Internet services, Making movies.
4. **Computer Networks:** Computer networks – basic principles of construction and operating, Networks as communication media, Network security, Wireless networks, LAN and WAN.
5. **New Tendencies in Informatics and its Applications:** Algorithms of the Internet, Can computers make business, Concurrency in informatics and in our life, Data exploration, JavaScript, Is P = NP or how to win million dollars in Sudoku, Enigma and contemporary cryptography, Past and the future of informatics – elements of history of informatics, Logic and computers, Introduction to neural networks, Medical informatics.

The courses are prepared and run by teachers from WWSI and from other universities in Warsaw and in Poland. There are more than 60 courses offered. For each course its authors prepare handouts (from 15 pages for a lecture to 50 pages for a 24 hour workshop), Power Point presentation (used during a lecture part of the course), tests and some other materials for students.

Two books will be published: "How to work with students talented in informatics – a guide for teachers" and "Homo informaticus – introduction to contemporary informatics" – it will consists of elementary introductions to various branches of informatics.

First year of the Project

The first year of the project in the school year 2009/2010 appeared to be our great success – 5500 school students participated in the courses: 2214 students – in 88 lectures and workshops (in WWSI), 412 – in 20 afternoon lectures (in WWSI), 2329 – in 47 lectures in schools, 450 – in 37 workshops (24 hours).

In Table 1 we present the results of questionnaires filled in by all participants of the courses, whereas Table 2 contains the results of questionnaires filled in by the students who graduated from vocational schools in June 2010.

The organizers of the project are very satisfied with the students´ opinions about the project, especially with the impact of the project on students' learning and on their positive attitude toward our proposals of courses and other activities.

Table 1. Students' opinion toward usefulness of the courses

Question	Yes definitely	Yes	No	No definitely	No answer
1. Are you interested in studying informatics in the future?	33%	33%	24%	10%	0%
2. Do you think that participation in the project will influence your future decision about your career?	22%	42%	26%	9%	1%
3. Has the course improved your knowledge and skills in informatics?	41%	47%	9%	3%	0%
4. Has the course encouraged you to develop your knowledge and skills in informatics by yourself?	26%	49%	22%	3%	0%
5. Have the materials been useful in the course?	55%	36%	7%	2%	0%

Table 2. Students' opinion toward influence of the courses

Question	Yes	No	NA
1. After taking part in the project, have you improved your grades in informatics?	46%	33%	21%
2. After taking part in the project, have you improved your grades in information technology?	46%	31%	23%
3. Has your choice of informatics related study been influence by the project?	62%	38%	–

Extensive workshops have been organized for contestants of the Olympiad in Informatics 2010. Twenty of them successfully reached the third final stage and two of them will represent Poland in the International Olympiad in Informatics in 2011.

Almost 1000 students participated in the Beaver contest. 5 students won the II Prize and 8 students won the III Prize.

Reflections

As the coordinator of the project I must admit that I am very satisfied with running the project, the enthusiasm of school students and teachers about our offer of courses and activities and the project's impact on schools – they are really interested in improving instruction in schools and giving students new opportunity to learn and develop their skills in the area of informatics curriculum topics and applications.

Personally, let me share one of my experiences. I run some of the algorithmic courses. Once, a group of young girls attended a course on introductory algorithmics. When before the workshop I learnt from their teacher that they have no experience in programming, I thought I would be in trouble but finally those girls were able to understand three simple algorithmic situations (e.g. for given three numbers interpreted as the lengths of triangle sides, find the area of the triangle if it exists) and write in Pascal and run successfully three programs. As one of my colleagues put it: everybody can be taught programming – now I strongly believe him. In fact, computer programming (in any sense) is a tool of computational thinking and as such should be a competence of everyone.

Conclusions

In conclusion, Informatyka + is a valuable project supporting the learning process in informatics and in information and communication technology in schools and helping students to choose their future career.

In the near future we intend to:

- apply to the Ministry of National Education to extend the project to all regions of Poland;
- make the project activities permanent and continues in schools;
- extend the scope of the project by constantly adding new topics, courses, activities.

5 Conclusions

We presented a number of activities in Poland which are outreach projects run nationwide and/or locally. We expect and have gathered some evidence that these activities increase motivation and preparation of school students for their future decisions to study computer science or related fields and become computer specialists.

The approach which we use can be viewed as implementation of computational thinking to teaching and learning informatics (computer science) topics and applications of computing in various areas of students' interests.

References

1. Bebras: International, http://bebras.org/en/welcome; (in Poland), http://www.bobr.edu.pl/
2. Computational thinking, http://www.iste.org/standards/computational-thinking.aspx
3. CSTA: National Secondary Computer Science Survey (2009), http://csta.acm.org/Research/sub/CSTAResearch.html
4. CSTA K-12 Computer Science Standards (2011), http://csta.acm.org/Research/sub/CSTAResearch.html
5. Denning, P.J.: Who Are We? Comm. ACM 44, 15–19 (2001)
6. Diks, K., Madey, J.: From Top Coders to Top IT Professionals. In: Mittermeir, R.T., Sysło, M.M. (eds.) ISSEP 2008. LNCS, vol. 5090, pp. 31–40. Springer, Heidelberg (2008)
7. Fronter, http://webfronter.com/iplus/milacollegejunior/
8. Gurbiel, E., Hardt-Olejniczak, G., Kołczyk, E., Krupicka, H., Sysło, M.M.: Informatics. Textbook for middle school, WSiP, Warszawa (2009) (in Polish)
9. Gurbiel, E., Hardt-Olejniczak, G., Kołczyk, E., Krupicka, H., Sysło, M.M.: Informatics for All Students. Textbook for high school, WSiP, Warszawa (2012) (in Polish) (in preparation)

10. Gurbiel, E., Hard-Olejniczak, G., Kołczyk, E., Krupicka, H., Sysło, M.M.: Informatics, WSiP, Warszawa. Textbook for high school, vol. 1, 2 (2012) (in Polish) (in preparation)
11. Informatyka +, http://www.informatykaplus.edu.pl/infp.php/
12. NSF, CS/10,000 Project, http://www.computingportal.org/cs10k
13. Olympiad in Informatics (in Poland), http://www.oi.edu.pl/, International, http://www.ioinformatics.org/
14. Running On Empty: The Failure to Teach K-12 Computer Science in The Digital Age, ACM, CSTA (2010), http://csta.acm.org/Runninonempty/
15. STEM: STEM Education Coalition, http://www.stemedcoalition.org/
16. Stephenson, C., Gal-Ezer, J., Haberman, B., Verno, A.: The New Education Imperative: Improving High School Computer Science Education, Final Report of the CSTA Curriculum Improvement Task Force, CSTA, ACM (February 2005), http://csta.acm.org/Publications/White_Paper07_06.pdf
17. Sysło, M.M., Kwiatkowska, A.B.: Informatics versus information technology – how much informatics is needed to use information technology – a school perspective. In: Mittermeir, R.T. (ed.) ISSEP 2005. LNCS, vol. 3422, pp. 178–188. Springer, Heidelberg (2005)
18. Sysło, M.M., Kwiatkowska, A.B.: The Challenging face of informatics education in Poland. In: Mittermeir, R.T., Sysło, M.M. (eds.) ISSEP 2008. LNCS, vol. 5090, pp. 1–18. Springer, Heidelberg (2008) (in Poland)
19. Wing, J.M.: Computational thinking. Comm. ACM 49, 33–35 (2006)
20. WWSI, http://www.wwsi.edu.pl, http://wscs.eu
21. Youth Academy of Informatics, http://www.main.edu.pl

Overcoming Obstacles to CS Education by Using Non-programming Outreach Programmes

Tim Bell[1], Paul Curzon[2], Quintin Cutts[3],
Valentina Dagienė[4], and Bruria Haberman[5]

[1] University of Canterbury, Christchurch 8041, NZ
tim.bell@canterbury.ac.nz
[2] Queen Mary University of London, London, E1 4NS, UK
paul.curzon@eecs.qmul.ac.uk
[3] University of Glasgow, Glasgow, G12 8RZ, Scotland
quintin.cutts@glasgow.ac.uk
[4] Vilnius University, Faculty of Mathematics and Informatics,
Naugarduko str. 24, Vilnius LT-03223, Lithuania
valentina.dagiene@mif.vu.lt
[5] Holon Institute of Technology, Holon,
Israel, and Davidson Institute of Science Education,
Weizmann Institute of Science, Rehovot 76100, Israel
bruria.haberman@weizmann.ac.il

Abstract. Formal Computer Science curricula in schools are currently in a state of flux, yet there is an urgency to have school students exposed to CS concepts so that they can make informed decisions about career paths. An effective way to address this is through outreach programmes that can operate outside or in conjunction with the formal education system. We compare 5 successful programmes. Each downplays programming as a pre-requisite skill for engaging with Computer Science ideas. This makes them accessible in short bursts without formal curriculum support. The formats used include contests, shows, magazine articles, and resources for teachers. We compare the 5 approaches to draw out key ideas for successfully addressing a school student audience. This can be used as the basis for designing new outreach programs.

Keywords: CS Education, Informatics Education, K-12 outreach, Information Technology, Computational Thinking.

1   Introduction

While formal school curricula around the world are gradually introducing Computer Science (CS) as a subject, there is an urgency to get school-age students interested in the topic, and consequently many outreach programmes have emerged that either work outside the school system, or supplement what is available in schools. Here we compare five such approaches to introducing Computer Science to high school age students. An important common feature of these is that none assumes programming as a preliminary or pre-requisite topic – they enable students to engage with concepts

I. Kalaš and R.T. Mittermeir (Eds.): ISSEP 2011, LNCS 7013, pp. 71–81, 2011.
© Springer-Verlag Berlin Heidelberg 2011

from Computer Science without having to first learn how to program. In a formal school programme there would be time to develop programming skills and provide good computing resources, but outreach programmes must necessarily make do with limited time with students and whatever facilities happen to be available.

Teaching CS without programming is achieved in a variety of ways including off-line kinesthetic activities, problem solving challenges that involve computational thinking, engaging magazine articles that present ideas from Computer Science, and mentoring by experts. All the approaches described here started as extracurricular outreach initiatives independent of formal school curriculum and constraints, thus there was a lot of freedom and space for imagination in their design. They also provide "grass-roots" trials of approaches for introducing students to CS, and the successful elements of these "trials" can later be absorbed into formal curricula. The approaches discussed here have all had widespread adoption and influence (typically tens of thousands of students) so there is potentially much to be learned from exploring their commonalities and differences. Our aim for this comparison is to compare and contrast existing successful approaches, drawing out common features and themes to give guidelines for the design of future initiatives.

There are three main motivations for downplaying the role of programming when first introducing students to CS. (1) In the context of outreach where a relatively short time is available to interact with students (often just a single short lesson), there isn't time to teach programming. (2) By engaging students without requiring them to learn to program first, a potential barrier is removed that could deter students from pursuing Computer Science. Some students will enjoy CS concepts such as problem-solving more than programming, and by engaging with those concepts first they will have more motivation to learn programming, which they are likely to encounter as a prerequisite for studying Computer Science formally. (3) A non-programming approach is a practical way to engage students in Computational Thinking [13]. Students thus gain benefits beyond a computing career.

In all the approaches described here, the key is that programming isn't the central element. Eventually students will likely need to learn programming, but it can be *after* students become engaged.

In the following sections we first describe each programme. In Section 7 we then provide a classification tool to highlight their similarities and differences which can be used to provide guidance on building a successful programme.

2 Bebras

Many competitions in computing and IT are intended for very talented students and focus on areas such as developing algorithms and programming. The Bebras competition instead has students solve problems from a broad range of areas without programming [6, 8, 9]. Because many students enjoy competition, such contests at school can be used to attract students to the domain covered by the contest.

The idea of a competition based around informatics and computer fluency for a wide population of high-school students started in Lithuania in 2003. It was named "Bebras" ("Beaver" in English) after the hard-working, persistent, intelligent, and lively animal. The main goals of the project are to promote students' interest in informatics (i.e. Computer Science) and Information and Communication Technology (ICT) from the start of their school career, to motivate students to learn and master computers, and to engage in computational thinking [6, 7, 13]. The contest is for all lower and upper secondary school pupils, divided into four age groups. Students have to solve 18 to 27 tasks on different levels within 45-60 minutes, entering answers via computer. They do not require prior topic knowledge, but do require students to be able to reason with common structures in the CS/informatics canon.

The tasks involve concepts such as algorithms (sequential and concurrent); data structures (heaps, stacks and queues, trees, and graphs); modeling of states, control flow and data flow; human-computer interaction; and graphics. Students do not study these topics formally, instead, the topics are introduced implicitly by having the students attempt imaginative tasks. A "narrative cover story" is used to relate the tasks to an underlying topic.

More than 10 countries now participate in Bebras. Since the contest is now international, one specific challenge is to find a balance between national and global standards for the contest. Hence, discussion on common standards and tasks suitable for all countries takes place at annual international workshops. A shared collection of tasks is developed including mandatory tasks to be included by all countries in their contests; additional tasks can be added to this to adapt the competition to the educational framework of each country. Surveys and informal feedback reported by the organizers in different countries suggest that the contest motivated students to get to know computer science and information technology better.

3 CS Unplugged

CS Unplugged [2] provides a variety of resources that engage students in Computer Science activities without using a computer. Instead of programming ideas from Computer Science, students interact with them through magic tricks, games and puzzles. Most activities have a strong kinaesthetic component and take a constructivist approach: students are given enough clues so they can work out principles themselves. A constructivist approach, where students are guided with leading questions so that they can discover CS principles for themselves, is important for outreach as it demonstrates to students that they could have invented much of the knowledge themselves; and of course, this is a lot more engaging than simply being told impressive facts.

The CS Unplugged resources are available for free download; as well as activities with specific guides for the presenter, there are videos demonstrating the activities or providing challenges to students, and links to extensive related material for follow-up. The activities have been translated into over a dozen languages. Another format for the material is a one-hour show [1], designed for a broad coverage of CS topics in a short time, rather than in-depth work by the students.

CS Unplugged is used in many situations around the world, generally relating to one-off events and visits, but increasingly as components of teaching programmes. It started around 1992 as a collection of ideas for outreach from universities to K-12 schools. Through its inclusion in the ACM K-12 curriculum in 2003 it started to be seen as the basis of a new approach for teaching CS in schools, either as the main material, or a supplement providing a break from being in a computer lab. It is now used widely in a variety of situations, including outreach, clubs, summer camps, and regular classrooms.

4 cs4fn

cs4fn [4] uses research topics to spark enthusiasm in students about Computer Science, providing them with ways to learn more about the subject so that the initial spark develops into a more sustained interest. It consists of four elements: a 20-page free magazine sent twice yearly to UK schools and subscribers worldwide; a website with up to 10 new articles per month; shows, such as the hour long cs4fn magic show [3], and booklets that allow audiences to explore topics more deeply.

The core target audience for cs4fn resources are school students aged 14+ though some are also presented to younger (9+) children and family audiences at science festivals. The magazines and website are read by people of all ages and professions including students, teachers, professionals and interested members of the general public.

The time commitment for cs4fn participants need not be high; school talks last only one school period, and at science festivals contact with individuals may be as short as 10 minutes. Likewise, students may only read a few articles at a time. The programme focuses on telling engaging stories about research, illustrating CS concepts using examples that resonate with students' lives. It also examines thought-provoking topics with a philosophical dimension, such as artificial intelligence. cs4fn talks incorporate lots of interaction. The project uses a concept of a 'sticky web' of approaches, the idea being that whether individuals find it via web, magazine or a talk, they are then drawn further in to the other strands of the project.

Across all strands cs4fn aims to show that computing is a fun, enjoyable subject that students should take further for intrinsic motivational reasons. Feedback from teachers and participants is overwhelmingly positive for all four elements of the project.

5 CS Inside

The CS Inside approach [5] draws much from CS Unplugged: predominantly kinaesthetic activities, undertaken without the need for computers, designed to be used by presenters with varying levels of computing experience. Whereas CS Unplugged was originally designed for students up to about the age of 12, the CS Inside activities were written for high school students from the start, recognising the different motivations that these two student groups will have for taking part in kinaesthetic activities. "Inside" is used in the approach's name because each activity brings out some of the Computer Science to be found inside the technology that is part of students' everyday lives, such as mobile phones, web browsers and game consoles. The aim is to engage students in issues of relevance to them about computing technology, and open up those issues by exploring the computer science inside.

The areas of Computer Science covered by the activities are not explicitly chosen to match with any particular school curriculum. The original context, however, was Scottish schools, and links are identified from the CS Inside activities to precise parts of the Scottish schools' computing curriculum where they may be useful. Some teachers use only those activities that help support the curriculum while others are happy to use them all (and are hungry for more!), recognising that materials of this nature are genuinely inspiring for students.

The activities are typically structured in four parts: the *Grab* captures the students' attention by asking questions about technology that are relevant to students; the *Intro* shifts attention from the students' context to the technology context to be addressed in this activity – this should be as small as possible to avoid losing the students' interest; the *Activity* is the main task; and the *Sustain* carries the learning from this activity out into their everyday lives, so that everyday events concerning technology will remind them of what is going on inside.

6 CS, Academia and Industry

This programme aims to expose students directly to state-of-the-art research, advanced technologies, software engineering methodologies, and professional norms by having the students interact with leading experts [14]. It is extracurricular, designed especially for talented high-school students in Israel who major in CS. Its main goal is to bridge the gap between school education and the "real world" of computing, especially relating to content, learning culture, and professional norms.

There are three main motivations for the approach: students can (1) add state-of-the-art computing research and development to the fundamentals taught at school, (2) be encouraged to become self-learners by experiencing a "taste-based", breadth-oriented learning approach, and (3) participate in "real-world" software development through a comprehensive project.

The two-year programme blends formal and informal learning and includes enrichment meetings, field trips and software development projects under the supervision of experts. Talented students are recommended to attend by their teachers.

The first stage is designed for 11th grade students and consists of a 7-month enrichment workshop looking at contemporary issues in computing. In the second phase, the 12th grade students (chosen from the first stage attendees) develop comprehensive software projects under the apprenticeship-based supervision of professional mentors (scientists and engineers from academia and the hi-tech industry).

An underlying principle of the first phase, which avoids using programming, is that students should be taught to employ a breadth-oriented learning style, in which their initial exposure to an unfamiliar topic will be accomplished by exposing them only to its essence (i.e., the main high-level abstract ideas). In other words, a complete understanding, including knowing the concrete details and mastering procedural aspects, should not be considered as the immediate aim of an initial exposure to a new topic. To achieve this, monthly enrichment meetings are conducted in which a variety of advanced topics are introduced in plenary sessions by leading representatives of CS/SE academia and industry. The sessions cover topics such as computer sciences and biology, artificial intelligence, computing in space, and professional norms.

In addition, the following non-programming learning activities were conducted, challenging algorithmic problems, role-playing simulation games, creative thinking in computer science, model-based-development, and a competition in testing software.

Feedback so far indicates that the programme contributes to developing a culture of learning befitting the dynamic world of industrial computing, thus providing the students with an entry point into the computing community of practice [10,14].

7 Comparison of the Programmes

Tables 1 and 2 compare the five approaches using criteria that highlight the similarities and differences between them. These criteria have emerged as a result of the previous discussion. By drawing out the commonalities of such large-scale successful programmes, we can draw lessons about important ingredients for future initiatives.

Table 1 focuses on the design of each approach. Comparing the five approaches, we see that in terms of the range of topics covered, all offer a breadth across Computer Science, exposing students to the range of topics that they might choose from to specialise in if they undertake the discipline. All operate on a large scale, with thousands of participants, and dozens (if not hundreds) of resources for presenters to draw on. All have a high level of visibility, some more locally, while others have a large international following. All rely on a pool of contributors and have some form of quality control. Most of the approaches engage students for around an hour at a time, though magazines and videos may only require minutes to engage with.

Table 2 analyses how the participants (students and teachers) interact with each programme. Finding a motivation for students to participate is key. Because the work is largely outside the curriculum, and therefore does not count towards formal grades, there need to be other motivations, either intrinsic or extrinsic. From the table, we can see that intrinsic motivation in the form of satisfying curiosity and enjoying problem solving are common to all the programmes. However, specific motivations used depend on the approach. A magazine article needs to capture enough interest at the start to keep the students reading for a few minutes. Humour and story-telling are used in the shows to keep a larger and possibly reluctant audience engaged for longer.

Programmes that require more commitment from students use prizes or certificates, and experts can serve as role-models to keep students engaged. A programme over several years needs either to use a deep intrinsic motivation of enjoyment in the subject or help students achieve their long term goals using, e.g., long-term career prospects as motivation.

All of the approaches considered avoid having programming as a primary focus, at least initially. The high level of uptake of all 5 indicates that this doesn't prevent students from being interested. These contrast with other popular approaches to outreach that are largely based around programming, such as robotics competitions, or introductory languages such as Scratch, Alice and Greenfoot [12]. The two-stage "CS, Academia & Industry" programme combines both approaches; it doesn't require programming in the preliminary stage, but the advanced stage is based around it.

We note that programming can be integrated in a variety of ways with these predominantly non-programming approaches, either as a follow-up where students implement ideas they have been exploring, or conversely where the non-programming

Table 1. A comparison of the five outreach approaches: design

	Bebras	CS Unplugged	cs4fn	CS Inside	CS, academia & industry
Web site	www.bebras.org; www.bebras.lt	csunplugged.org	cs4fn.org	csi.dcs.gla.ac.uk	davidson.weizmann.ac.il/eng/projects.php?cat=488
Topics covered	Broad coverage of technical and soft topics	Broad coverage of technical topics	Research based, technical, soft and interdisciplinary topics	Broad coverage of technical topics	Broad coverage of state-of-the-art R&D in various CS and software engineering topics
Scale of resources	100-200 tasks each year (since 2004)	25 main activities, 49 videos, more added each year	600— articles, 6-10 new per month, 10 talks per month	11 main activities. Further activities developed each year	A collection of projects developed by the students
Scale of activity	200,000 participants in annual competition	850 website page views daily, 100 video views daily	Annually 2-3 magazines, up to 23,000 copies, 70 shows to 9000 students. 2000 website visitors daily.	1200 registered users. 250 Scottish schools. 700 activity downloads per year.	Over 7 years, >1,000 total students attended first stage, 200 of whom developed projects in the advanced stage
Adoption	20 countries are involved (2010), tasks translated into 15 languages	International, translated into 14 languages, in ACM K-12 curriculum	International website and magazine readers. Some material translated into 4 languages	Adopted principally in UK, with individual users from over 20 countries.	Conducted at one site, attended by students from 30 schools across Israel.
Contributors to resources	Annual workshop from multiple countries. Teachers and CS students contribute.	Suggestions from around the world, collected and quality controlled by project leader	Academics and team at Queen Mary, University of London with article contributions mainly from UK	Academics and students at University of Glasgow	Academics at Weizmann Institute of Science and other universities in Israel plus industry mentors
Typical time per activity	Contest: 45-60 mins. Some countries have two-round contests.	Classroom: 1 to 2 hours per activity; Show: 1 hour; Videos: a few mins	Shows: 45-120 mins; Festivals: typically 10 minutes contact; Articles: 10 mins	30-70 minutes per activity	Preliminary stage: 50 min lecture <> 2 + 90 min activity monthly; Advanced stage: over 9-10 months.

Table 2. A comparison of the five outreach approaches: analysis

	Bebras	CS Unplugged	cs4fn	CS Inside	CS, academia & industry
Student motivation	Interesting, attractive tasks, problem solving, certificates, prizes, shows	Kinaesthetic activity, problem solving, curiosity, humour, small prizes	Curiosity, offbeat links, humour, engaging stories, kinaesthetic activities	Relevance to technology in everyday lives, kinaesthetic, problem solving, curiosity, school testing	Intrinsic interest, career prospects
Prerequisites	Basic mathematical and reasoning skills	Basic mathematical and reasoning skills	None for most, some require basic reasoning skills	Basic mathematical and reasoning skills	Studying computer science according to a formal national program
Relationship to programming	No programming skills required. Logo-based commands used in some problems.	No programming, but can use programming as follow-up, or use activities as break from programming.	No programming, though some articles explain programming concepts.	Principally no programming. Ideas can be programmed as a follow-up.	First phase: no programming Second phase: programming project
Relationship to curriculum	Intends to influence common understanding of CS curriculum for high schools. A subset of problems may relate to national curricula.	Usually extra-curricula, but links to curriculum provided and can be used to supplement curricula material.	Does not address curricula directly, but goes beyond it. Teachers use it as support to the curriculum with some guidance.	Some activities designed to support Scottish curriculum. Links to Scottish curricula included.	Supplement to school curriculum.
Teachers' involvement	Teachers motivate students to attend run preparatory activities, and organise local online contest. At end link the tasks to informatics.	The material can be presented by teachers, or they organise visits from/to universities	The material can be presented by teachers, or they organise visits from/to universities	The activities are primarily run by teachers, but students and academics present them too	Teachers accompany students through the whole 2-year period; attend enrichment meetings and mediate between the students and their mentors throughout project

activities provide a physical break from programming at a computer. In fact, there is evidence that including programming as a follow-up is useful, to help students see more deeply how the material relates to computers [11].

The material discussed in this paper is not generally taught as part of a formal curriculum, typically being used in outreach and interest programmes such as one-off visits, out-of-school clubs, science centres, and special events. This is particularly useful if CS isn't compulsory in schools, as is the case in many countries and states, as it helps to expose students to a subject that they may have little idea about. It also helps teachers and careers advisers to understand the topic, supporting them to help their students make appropriate career decisions.

Looking at the impact on teachers, we note that the programmes support teachers by providing resources for them rather than expecting them to do the preparation and planning. In fact, an unexpected focus of the CS Inside project was the development and support of teacher communities. Simply providing materials to teachers isn't sufficient; high school CS teachers are often a forgotten community, and making personal contact and building trust has been a key component of CS Inside's success. The other approaches also report spin-offs from an improvement in relationships with teachers, partly by providing a reason for contact between schools and universities, and also in helping teachers gain a significantly better understanding of CS concepts.

Each of these approaches either already shares ideas with the others, or can benefit from sharing them. For example, some of the CS Inside and cs4fn lessons are based on CS Unplugged activities; conversely, the CS Unplugged website provides links to the relevant CS Inside and cs4fn activities as extension or follow-up ideas. New puzzles in the Bebras project can be constructed by looking at the challenges in the other projects and creating a story/context to make them accessible to contestants, and phase 1 of the CS, academia and industry program can use activities and lesson plans from the other approaches to provide a means to engage with students. The non-programming approaches here also work well as supplements to more traditional programming-based approaches, providing an active break from sedentary work at the computer.

All the approaches are flexible – the resources themselves generally provide creative ideas for teaching that can be adapted to the audience and by the audience (teachers or students) for their own needs. The programmes are presented in a way that is accessible for teachers and other organisers; resources are provided at no direct cost through well-resourced websites, and the details that can be time-consuming for teachers (such as preparing slides and handouts) are taken care of. Notably, all the programmes include significant commitment and resources over a long period of time from their organisers.

8 Conclusions

There are a variety of ways that students can be exposed to Computer Science without the barrier of requiring programming as a pre-requisite. The large number of students who have participated in the approaches described here indicates that the avoidance of programming can indeed generate ongoing interest and so can be considered successful. This contrasts with approaches where programming is taught in-depth first; in schools where time is limited the programming-first approach can lead to the misconception that CS is *only* about programming, and thus only students whose main interest in computing is programming are motivated to continue in CS.

Our analysis here shows that despite the five approaches having quite different formats, there are considerable commonalities between them that can guide the development of new initiatives. One valuable motivation for students to engage with these activities is a deeper understanding of the computing devices all around us. However, because the focus in all of these approaches is Computer Science, not computers, alternative motivators for students can be important: contest prizes, the challenge of solving a problem, curiosity, humour, and ideally, appealing to the intrinsic interest of the student in this kind of thinking and reasoning. This means that the material needs to be carefully crafted to attract and retain student interest (e.g. direct relevance to their life, engaging story telling, well planned magic tricks, questions to stimulate their curiosity), or it needs to create a culture that attracts students (e.g. past participants recommend it, high status or rewards for competition winners, a good reputation for the event).

The non-curriculum approach provides the opportunity for grass-roots influence on formal curricula when a top-down state-led approach often struggles to efficiently deliver CS education in schools. The success of several of the approaches here has led to them being recommended for curricula, so the authors then have an influence on formal CS education. An important issue that will need to be addressed then is how materials largely designed for outreach can be adapted for settings where a certain level of assessment will inevitably be required.

We have presented several creative approaches with the common goal of attracting students to study Computer Science. As a result of the comparison between them we developed the criteria presented in this paper. These criteria will enable those designing outreach and teaching programmes to evaluate approaches, to choose the most suitable approach for their students, and to adapt the approaches to the target population and context.

Acknowledgments. We are grateful to Cecile Yehezkel, Peter McOwan and Jonathan Black for their work with some of these programmes, and EPSRC, Google and our own institutions who have supported them.

References

1. Bell, T.: A low-cost high-impact computer science show for family audiences. In: Australasian Computer Science Conference, Canberra, Australia, pp. 10–16 (2000)
2. Bell, T., Alexander, J., Freeman, I., Grimley, M.: Computer Science Unplugged: School Students Doing Real Computing Without Computers. The New Zealand Journal of Applied Computing and Information Technology 13(1), 20–29 (2009)
3. Curzon, P., McOwan, P.W.: Engaging with Computer Science through Magic Shows. ACM SIGCSE Bulletin 40(3), 179–183 (2008)
4. Curzon, P., Black, J., Meagher, L.R., McOwan, P.W.: cs4fn.org: Enthusing Students about Computer Science. In: Proceedings of Informatics Education Europe IV, Freiburg, Germany, November 5-6, pp. 73–80 (2009)

5. Cutts, Q., Brown, M., Kemp, L., Matheson, C.: Enthusing and informing potential computer science students and their teachers. ACM SIGCSE Bulletin 39(3), 196–200 (2007)
6. Dagienė, V.: Information technology contests – introduction to computer science in an attractive way. Informatics in Education 5(1), 37–46 (2006)
7. Dagienė, V., Futschek, G.: Bebras International Contest on Informatics and Computer Literacy: Criteria for Good Tasks. In: Mittermeir, R.T., Sysło, M.M. (eds.) ISSEP 2008. LNCS, vol. 5090, pp. 19–30. Springer, Heidelberg (2008)
8. Dagiene, V.: Supporting computer science education through competitions. In: Proc. 9th WCCE 2009, Bento Goncalves, Paper-Nr. 76, 10 pages (2009)
9. Dagiene, V., Futschek, G.: Bebras International Contest on Informatics and Computer Literacy: A contest for all secondary school students to be more interested in Informatics and ICT concepts. In: Proc. 9th WCCE 2009, Bento Goncalves, Paper-Nr. 161, 2 pages (2009)
10. Haberman, B., Yehezkel, C.: A computer science educational program for establishing an entry point to the computing community of practice. J. of Information Technology Education (JIRE) 7, 81–100 (2008)
11. Taub, R., Ben-Ari, M., Armoni, M.: The effect of CS unplugged on middle-school students' views of CS. SIGCSE Bull. 41(3), 99–103 (2009)
12. Utting, I., Cooper, S., Kölling, M., Maloney, J., Resnick, M.: Alice, Greenfoot, and Scratch - A Discussion. Trans. Comput. Educ. 10(4), Article 17, 11 (2010)
13. Wing, J.M.: Computational thinking. Communications of the ACM 49(3), 33–35 (2006)
14. Yehezkel, C., Haberman, B.: Bridging the gap between school computing and the "real world". In: Mittermeir, R.T. (ed.) ISSEP 2006. LNCS, vol. 4226, pp. 38–47. Springer, Heidelberg (2006)

CS Unplugged Assisted by Digital Materials for Handicapped People at Schools

Hiroki Manabe[1,2], Susumu Kanemune[1], Mitaro Namiki[3], and Yoshiaki Nakano[1]

[1] Osaka Electro-Communication University, Japan
[2] Hadano-Sogo High School, Japan
[3] Tokyo University of Agriculture and Technology, Japan
manaty2005@mh.scn-net.ne.jp,
kanemune@acm.org,
namiki@cc.tuat.ac.jp, info@nakano.ac

Abstract. We report practice lessons in 'Computer Science Unplugged' (CS Unplugged) with assisted digital materials. CS Unplugged involves physical or group activities that lead students to computer science, and it is an excellent method of informatics education for beginners. However, such activities are not always easy for all students. Therefore, we designed various digital materials to assist handicapped students with such activities. We also adopted these to the lessons of CS Unplugged at a vocational training school for the disabled. As a result, we observed that the materials effectively assisted students with their lessons.

Keywords: Computer Science, computer science unplugged, informatics education.

1 Introduction

Computer Science Unplugged (CS Unplugged) [1][2] is an excellent method of learning the basics of computer science. There are many junior/senior high schools and universities in Japan in which CS Unplugged has been used [3]. CS Unplugged has three outstanding features [4].

1. It consists of active games (e.g., drawing/painting, magic tricks, and group learning).
2. Each activity leads to students learning some concepts of computer science.
3. None of the activities require computers.

The authors adopted CS Unplugged in the curriculum for a course at a vocational training school for people with disabilities in 2008 and tested and confirmed its learning effects. However, we noticed that kinesthetic activities might not be able to be exercised in class because of physical or communication problems.

This paper discusses a solution that is assisted by computers to CS Unplugged activities. The authors developed various digital materials to solve these problems. We report practice lessons and the effect of these materials.

I. Kalaš and R.T. Mittermeir (Eds.): ISSEP 2011, LNCS 7013, pp. 82–93, 2011.
© Springer-Verlag Berlin Heidelberg 2011

2   Adoption of CS Unplugged in Vocational Training School for Handicapped People

2.1   CS Unplugged Content

One of the authors translated CS Unplugged into Japanese and had 'Informatics not using computers' published in 2007 [5][6]. The book detailed 12 activities. Table 1 summarizes its contents.

Table 1. Informatics not using computers

Activity	Title	Activity's Content	Digital Materials
1	Count the Dots	*Binary Numbers*	✓
2	Color by Numbers	*Image Representation*	✓
3	You Can Say That Again!	*Text Compression*	
4	Card Flip Magic	*Error Detection & Correction*	✓
5	Twenty Guesses	*Information Theory*	
6	Battleships	*Searching Algorithms*	
7	Lightest and Heaviest	*Sorting Algorithms*	
8	Beat the Clock	*Sorting Networks*	✓
9	The Muddy City	*Minimal Spanning Trees*	
10	The Orange Game	*Routing and Deadlock in Networks*	✓
11	Treasure Hunt	*FiniteState Automata*	✓
12	Marching Orders	*Programming Language*	

For example, Activity 4 (Card Flip Magic) treats error correction by using parity bits. This lesson begins with a demonstration of a card magic trick by the teacher.

First, a student places many two-sided cards on a blackboard as a matrix of 5x5 squares. The teacher (magician) places more cards in a row and in a column while saying 'this is just to make it a bit harder'. Next, the teacher turns around, does not face the cards, and the student flips over one of the cards. Last, the teacher can hit the card, which the student flipped over, even though the teacher had been looking in the opposite direction. After the magic trick, the teacher lets students think about why he/she could hit the card, where additional cards should be placed, and other details. These questions lead to the concept of parity bits. If students notice or discover this concept by themselves, they gain greater educational benefits rather than just being taught by the teacher.

Thus, CS Unplugged is a learning method that raises students' motivation and attracts them to the world of computer science.

2.2   Adopting CS Unplugged to Vocational Training of Disabled Students

In 2008, we tried to adopt CS Unplugged in the curriculum for the 'OA System Course' of Kanagawa Vocational Training School for students with disabilities. This school was established with the aim of supporting disabled people gaining social independence through vocational capabilities. Computer literacy was the most important skill they had to acquire.

The OA System Course, which is for physical disabilities, is a special course that grooms students to become computer programmers or systems engineers. Therefore, the students have to study computer science and computer technology. The main content of vocational training, which is related to computer technology, is generally to teach how computer programs are made. Therefore, traditional learning content and learning methods were adopted for this course.

However, many of the students are enrolled at this school with the aim that they could be rehabilitated into society after having overcome difficulties they suffered past accidents or sickness. They wanted to make information processing their occupation, but they had no interest in programming and disliked thinking logically. Additionally, there was one student who was not able to sufficiently learn through the process of compulsory education due to long-term hospitalization and there was another who was not able to take notes due to physical reasons. Therefore, even a book for beginners was sufficiently obscure to decrease their motivation to learn. The conventional learning approach was not suitable for basic study of their future occupations.

Therefore, we decided to try to adopt CS Unplugged into the curriculum because it does not require assumed knowledge and it leads to the basics of computer science. Therefore, all students could learn about computer science without difficulties. We also expected that CS Unplugged might develop their logical thinking abilities.

There had been no instances where CS Unplugged had been adopted in any vocational training curricula. First, students seemed to be confused due to such a strange method of learning. However, they found CS Unplugged was an excellent learning method as they progressed through the curriculum. However, we confirmed that some features such as kinesthetic activities, physical movements, and communication skills in CS Unplugged became a serious issue for students with disabilities.

3   Issues and Achievements with CS Unplugged Practice for Students with Disabilities

3.1   Issues with CS Unplugged Practice for Disabled Students

We examined what kind of issues existed when 'normal' activities were carried out[7]. For example, students who had upper limb disorders could not understand activities such as 'coloring in with a pencil' or 'grabbing with their hands'[8][9]. These difficult activities affected the movements of moving materials in Activity 1 (Count the Dots), Activity 10 (The Orange Game), and the work time for 'coloring in with a pencil' in Activity 2 (Color by Numbers).

Physical movements in Activity 8 (Beat the Clock) or in Activity 11 (Treasure Hunt) were dangerous for students who had lower limb disorders. Consequently, such students could not participate in these activities. One student who had communication problems found it difficult to practice cooperative activities such as those in Activity 6 (Battleship) or Activity 10 (The Orange Game).

3.2   Development of Material Assisting Handicapped Students

One of the authors developed six online materials, which were based on CS Unplugged (checked in the fourth row of Table 1) to assist students with the activities.

The materials were uploaded to a Web site [10]. These materials were developed for personal use. There were three support patterns.

1. Replacing some actions (e.g., drawing/writing) by clicking with a pointing device.
2. Replacing group work with personal work.
3. Simulating group work.

The materials could be used repeatedly in the lesson or after it. Fig. 1 shows screen shots of the materials.

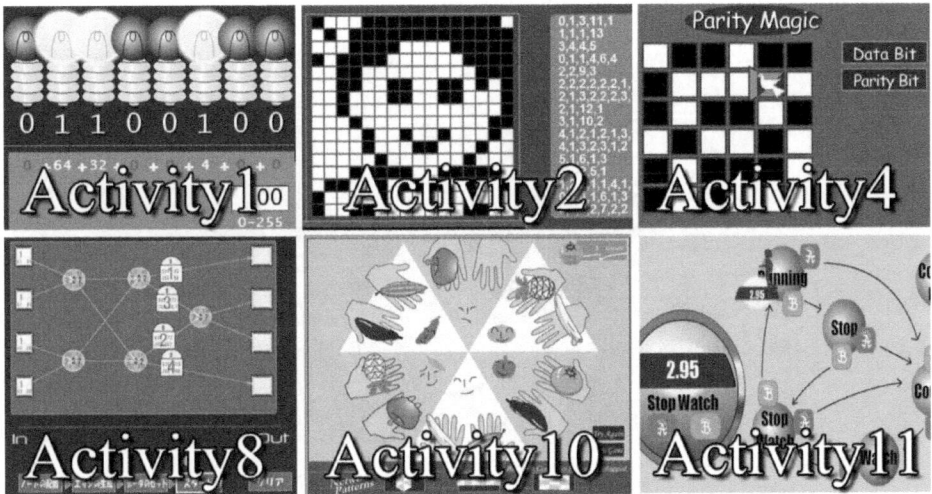

Fig. 1. Interfaces for CS Unplugged digital materials

The virtual world could also be an assisted environment. One of the authors had previously participated in research on CS Unplugged for disabled people [11], which was executed in the 3D virtual world of Second Life. Students in the virtual world could move around freely and communicate with one another by transforming themselves to avatars. Therefore, it was possible to practice CS Unplugged activities in this virtual world.

4   Trials on Digital Material to Provide Learning Support

4.1   Activity 2 (Color by Numbers)

This section introduces the lesson content for Activity 2 (Color by Numbers). This activity dealt with how images were digitalized to represent images on computers.

A run length algorithm was used to digitize the images for this activity. Students could learn the basics of digitalization and data representation of images throughout the work of coloring in dots in pencil, translating dots to numbers, and simulating communications. Our lesson plan involves seven steps.

Lesson plan:

1. The teacher urges students to notice the rules for the run length algorithm through a quiz game.
2. The students draw/paint an image on a card by coloring in small dots in pencil.
3. They translate/digitize the image to digital data with the run length algorithm.
4. Students exchange cards, which are written in digital data.
5. They color/encode the digital data that they received to dots. (Fig. 2)
6. They compare two images of the source and destination.
7. The teacher asks students how a computer stores data and what the important elements about network communication are.

Fig. 2. Activity 2 (Color by Numbers)

Most of the students in the class lesson were surprised at the principles underlying digitization or the communication of digital data. It was important for students to understand such principles in this lesson by coloring in the dots. Coloring work was done twice. This work was important to establish the relationship between colors and numbers. However, coloring in dots using a pencil is difficult for students who have upper limb disorders and erasing dots is actually more difficult than coloring them in. This caused such students to work inefficiently or lose concentration while learning.

We adopted digital materials based on Activity 2 in the lesson to solve these problems. These materials assisted students with manual coloring by enabling them to click with a mouse or use a track ball. These were used in lesson plans 2 and 5. The functions of these digital materials are explained below.

The canvas consisted of many small square tiles. The default number for the tiles was 256 (16x16). The students could change the tiles by clicking with the mouse. Each tile had two states, one side was white and the other was black. If a white tile was clicked with a pointing device, it converted to black and if a black tile was clicked, it converted to white.

Fig. 3 is a photograph of a student who had a severe impairment to his upper extremities. His grip was so weak that he could not use a pencil or eraser. He usually used a ballpoint pen for writing that was fixed to his hand with special equipment. He operated the track ball as a pointing device to use the computer and had two special pieces of equipment fixed to each hand to attach the ballpoint pen to and to type on the keyboard with the pens.

Fig. 3. Student with impairment using upper clicked trackball with special equipment for drawing/painting

The student spent too much time when he drew in pencils by hand and a trace protruded outside the frame. However, when he drew with the materials, he could reduce his working time. He was freed from having to worry about whether the trace had protruded outside the frame.

The student's four main comments were:

1. I felt good as I did not want to color in the dots in free hand drawing.
2. I wondered why the numbers became a painting.
3. I enjoyed this lesson even though my drawing was poor.
4. I understood the basics of digitization by using this learning material.

We found that the materials decreased the difference in working time between the student with upper limb disorders and other students. The student could also concentrate on his original studies without having to worry about whether the trace of the pen was outside the frame by using these materials.

4.2 Activity 10 (The Orange Game)

This section introduces the lesson for Activity 10 (The Orange Game). This activity dealt with the routing algorithm for a computer network. Students could learn about these concepts by delivering oranges (we used some plastic fruit) as packets.

This activity was usually practiced in groups and students could note the importance of processing efficiency through working collaboratively. Therefore, all communication skills by the participants were important factors for problem solving.

Our lesson plan involved five steps:

1. A group of six students sits in a circle.
2. The teacher distributes fruits to students randomly and each student has two pieces of fruit. (One student holds one piece of fruit with his/her other hand empty).
3. The teacher explains the rules of the game where students can only pass the fruit to both sides.
4. All students cooperate throughout the activity.
5. The teacher explains routing in a computer network after the game.

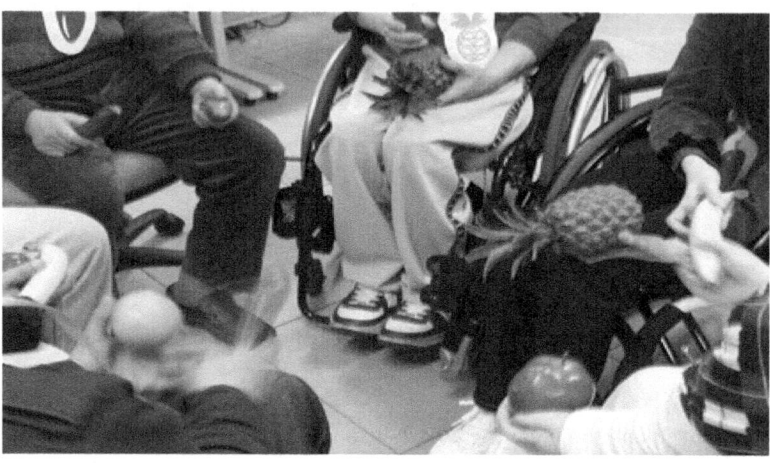

Fig. 4. Activity 10 (The Orange Game)

Group members in this activity had to pass the fruit to aim for the goal where all members were holding their own fruit. If any member thought that "I got my own fruit and I've finished", then the entire situation might not lead to a solution. It was important for each member to recognize the group status and share their method of delivering fruit to avoid such situations.

The conditions in classrooms at the vocational training school were as follows. The class for the lesson was divided into three groups. We observed the activities of all three groups. The first group evolved throughout the game by taking note of the time and counting the number of deliveries. The second group discussed how to obtain a better solution. However, the third group that was composed of three deaf students and three students with normal hearing did not communicate sufficiently and could not reach the goal even once. It was obvious that there was a lack of communication between the deaf students and those with normal hearing. We observed their efforts to communicate when they used gestures or writing. However, these were not sufficient to enable them to communicate effectively. This reduced their motivation and the activity was terminated without being completed. This meant they could not fully realize the importance of processing efficiency through collaborative work.

We predicted that if all members of the group attained a high level of ability in attaining a solution, then the whole group would reach the goal. We adopted digital materials based on Activity 10 in the lesson to develop personal abilities to solving this routing problem. The students played the orange game alone (Fig. 5) by using these materials. All the fruit had to be moved through individual thinking. Therefore, we expected individual abilities would increase. The fruit could only be placed on a neighboring player's hand in the same way as in the real game. The number of times fruit was received was counted. They could focus on learning objectives to simplify non-essential tasks, such as painting with colors.

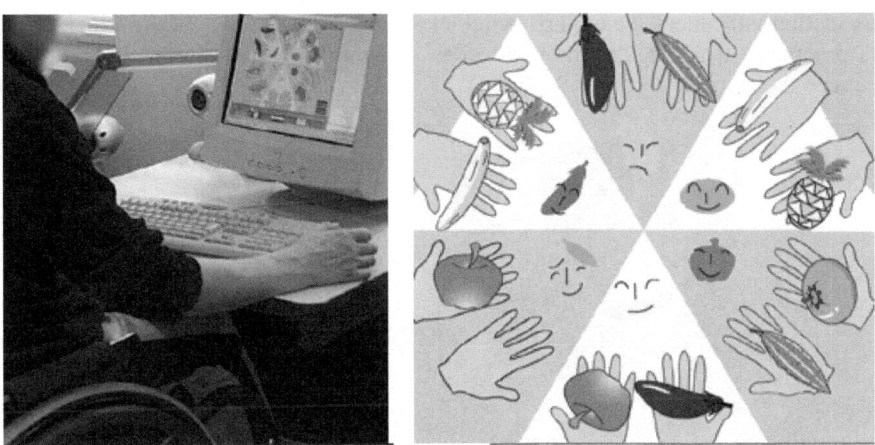

Fig. 5. Student in wheel chair (left) and material interface (right)

We taught an experimental lesson in another course at the vocational training school. First, the lesson began with the 'normal' orange game. Next, we made students use the materials. Last, the students played the 'normal' orange game again. Then, all groups were able to reach the goal.

Repeating the same games made some students get the knack of obtaining a solution. Moreover, we could see how students considered solving the problem by observing their computer displays. This enabled us to give appropriate advice to individual students.

The four main comments students made are below.

1. I felt it was hard because I had to move the oranges only by thinking.
2. I could do it well by myself but my group could not.
3. I felt the difference between the flat screen of the personal computer and the actual three-dimensional sensation was odd.
4. I understood I had been repeating useless movements by using these digital materials.

Three more tendencies were discovered.

1. Two opposite opinions coexisted. "My group could not do it well but I could " and "It was hard to think only by myself on the materials".
2. There were some opinions about the differences between the computer screen and the actual appearance of the fruit.
3. There were many opinions that required the best solution to be indicated in the materials.

We could infer the relationship between personal ideas and group work, which was not evidenced in the group activity by observing the students using the materials. Moreover, we confirmed that students adopting personal learning with the materials led all of them to think more deeply about the performed algorithm.

4.3   Activity 8 (Beat the Clock) in Second Life

This section introduces the experimental lesson for Activity 8 (Beat the Clock), which was practiced in a 3D virtual world. The students in this experiment had already practiced the 'normal' Activity 8 (Fig. 6).

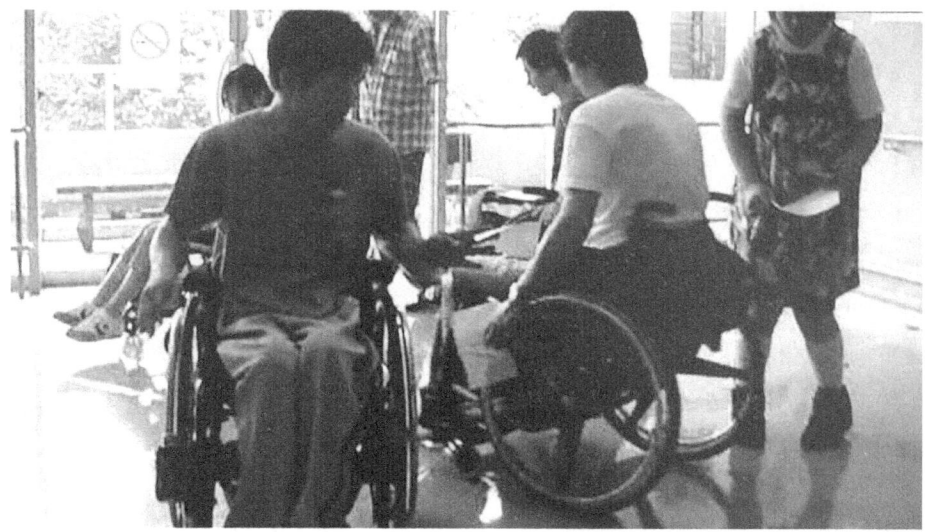

Fig. 6. Activity 8 (Beat the Clock)

This activity dealt with parallel computing where students could learn about this concept by walking on a parallel sorting network marked on the floor. The network was constructed with various lines and nodes. This activity was usually practiced by six students who walked along the lines of the network comparing the numbers on cards that they held at each node. When the students reached the goal, they could see that six numbers were perfectly sorted in order. This surprised the students and aroused their intellectual curiosity.

This activity was dangerous for students who had lower limb disorders. Some students did not participate in the 'normal' activity to avoid accidents.

However, even disabled students could move around freely in the virtual world by walking, running, jumping, and flying. A large network had been constructed in Second Life (Fig. 7) in this research.

First, the students tried to carry out 'Beat the Clock' in Second Life. They were able to execute this without any feelings of danger. Six large cubes and one button had been prepared in this sorting network area. Each cube had a number written on its surface. When an avatar pushed the button, the cubes began to move along the network lines to compare the numbers. The avatars and their users (students) could watch the movements of the cubes. Some avatars flew into the sky and experienced a bird's-eye view of all the movements.

They could not look over all the movements in the 'normal' activity. However, they could recognize what they had done from all viewpoints in the 'virtual' activity. This meant that virtual activities had other effects that real activities did not.

Most students enjoyed the experience and said they would like to use it more. The students who could not practice in the real world appreciated this implementation. We confirmed that the digital environment could remove factors that prevented learning with CS Unplugged in this research.

Fig. 7. Activity 8 (Beat the Clock) in Second Life

5   Using Digital Materials in High School Lessons

We recognized the effectiveness of the materials in alleviating physical problems when CS Unplugged was adopted for the practice lessons at the vocational training school. We also observed that the materials had effects other than assuaging their physical problems. Consequently, we used them in a high school lesson on a compulsory subject called 'Information'. We taught the lessons at Hadano-Sogo High School and observed the students.

5.1   Activity 2 (Color by Numbers)

We began to do 'normal' unplugged activities as we handed out paper and pens to the students. We found some students who concentrated on painting in dots or who spent too much time on painting in dots rather than coding data in the primary learning materials. Therefore, we taught other classes to use the digital materials where drawing/painting were alternated with clicking. As a result, these problems decreased and students focused on coding.

5.2 Activity 10 (Orange Game)

The lesson began with the 'normal' unplugged orange game, where the group was divided into two types of students, in which the first were directors and the second was directed. The directed students did not seem to think for themselves. Therefore, we made the students use the materials, where all students cooperated in all movements by group members. As a result, we found that all students began to think about the 'effective delivery of packets'.

5.3 Activity 8 (Beat the Clock)

We developed other digital materials (not Second Life) for Activity 8 (sorting network). We discovered that there were two problems with this activity. The first was that there were some students who could not understand what this activity meant. The second was that students could not watch all movements and they did not understand what was happening.

We implemented a bird's-eye view to solve these problems, where students could look down on the whole network. We also implemented 'changing the number of members, where they could increase/decrease the number (data). Students were able to think about the meaning of this activity and understand all movements by using the materials after the original activity.

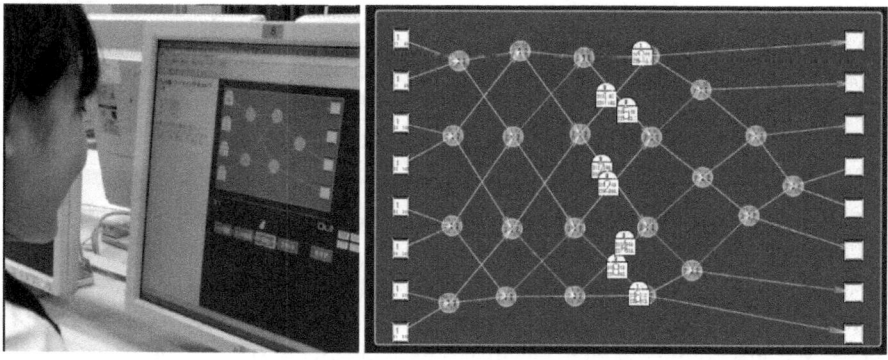

Fig. 8. Photo of Activity 8 (Beat the Clock) with high school student and screen shot

6 Conclusion

We tried to adopt CS Unplugged in the vocational training of disabled students and tried to support them by using digital teaching materials.

The students' learning attitudes were very positive in all activities and CS Unplugged was an appropriate learning method for vocational training for disabled students.

There were some activities in which disabled students could not participate thoroughly enough. However, handicapped people also have rights to experience excellent learning methods such as CS Unplugged. The digital materials supported them by giving them chances to gain a positive learning experience.

The students in general high school education who used the digital materials developed by one of the authors, demonstrated three different effects to the students with disabilities.

1. They could focus on learning objectives to simplify non-essential tasks, such as painting with colors.
2. They could think about steps in the algorithm by trial and error alone.
3. They could understand the activities by using the bird's-eye view of the entire activity.

We would like to develop digital teaching materials in the future that are more useful by listening to the opinions of numerous students and teachers.

References

1. Bell, T., Witten, I.H., Fellows, M.: Computer Science Unplugged: An enrichment and extension programme for primary-aged children, Lulu (2002)
2. Bell, T., Alexander, J., Freeman, I., Grimley, M.: Computer Science Unplugged: School students doing real computing without computers. The NZ Journal of Applied Computing and Information Technology 13(1), 20s (2009)
3. Nishida, T., Idosaka, Y., Hofuku, Y., Kanemune, S., Kuno, Y.: New Methodology of Information Education with Computer Science Unplugged. In: Mittermeir, R.T., Sysło, M.M. (eds.) ISSEP 2008. LNCS, vol. 5090, pp. 241–252. Springer, Heidelberg (2008)
4. Nishida, T., Kanemune, S., Namiki, M., Idosaka, Y., Bell, T., Kuno, Y.: A CS Unplugged Design Pattern. In: SIGCSE 2009 (2009)
5. Kanemune, S., Kuno, Y.: Informatics not using computer. Etext Laboratory (2007) (in Japanese)
6. Bell, T., Wada, T.B., Kanemune, S., Xia, X., Lee, W., Choi, S., Aspvall, B.: Making Computer Science activities accessible for the languages and cultures of Japan, Korea, China and Sweden. In: SIGCSE 2009, p. 566 (2008)
7. Lazzaro, J.J.: Adaptive Technologies for Learning & Work Environments. American Library Association (2001)
8. Turcsanyi-Szabo, M.: Designing Logo pedagogy for elementary education. In: EuroLogo 1997 (1997), http://eurologo.web.elte.hu/lectures/papthij.htm
9. Norte, S., Castilho, N., Condado, P.A., Lobo, F.G.: GoGoBoard and Logo programming for helping people with disabilities. In: EuroLogo 2005 (2005)
10. Manabe, H.: Information Classroom Near The Sea, http://www.info-study.net/ (in Japanese)
11. Bell, T., Grimley, M., Bianco, G., Marghitu, D., Manabe, H.: Kinesthetic Computer Science activities in a virtual world. In: SIGCSE 2009 (2009) (poster)

Computer Science in Primary Schools –
Not Possible, But Necessary?!

Ernestine Bischof and Barbara Sabitzer

Alpen-Adria Universiät Klagenfurt,
Institut für Informatiksysteme,
Universitätsstraße 65-67, 9020 Klagenfurt am Wörthersee, Austria
{ernestine,barbara}@isys.uni-klu.ac.at

Abstract. This paper reports on the main results obtained by an Austrian initiative attempting to broaden the pupils' view on Computer Science / Informatics and to create interest in the subject. Pupils from primary school up to upper secondary school obtained lectures by university teachers spread over a period of one and a half year. This paper presents one of the lectures and reports the results obtained in primary schools compared with results from other grades.

Keywords: CS unplugged, primary school, evaluation, view on CS.

1 Introduction

The project *Informatik erLeben* (experiencing informatics) aims at attracting students for Informatics as a constructive, technical discipline. However, as the concept of "Informatics in School" covers in many countries concepts ranging from introducing pupils into using information and communication technology (ICT) while the project reported upon is strictly concerned with constructive aspects and basic fundamental ideas Informatics contributes we rather use the term Computer Science (CS). This is to help avoiding confusion, since on the tertiary level, the terms Informatics and Computer Science are used interchangeably anyway. The lessons developed show pupils of all grades selected core-concepts of Informatics/CS in a playful way and at an age-specific level. The prepared lessons cover the topics shown in table 1 and partially presented already in [2].

In table 1, grey fields in table 1 show topics adequate for primary school pupils. All other topics can be taught to pupils at secondary schools. The table shows that the core-concepts (first column) are divided into several modules that can be composed individually. E.g., the network modules (6th row) can be combined with error detection (2nd row). The topics listed were developed in the course of more than a year and are continuously enhanced. In general modules are mutually independent. Some of them contain ideas from CS Unplugged [1]. Topics are selected in cooperation with the class teachers and presented in interventions[1] lasting about 90 minutes.

---

[1] We use the expressions lesson and intervention as synonym. They can contain one or more modules.

I. Kalaš and R.T. Mittermeir (Eds.): ISSEP 2011, LNCS 7013, pp. 94–105, 2011.
© Springer-Verlag Berlin Heidelberg 2011

Table 1. Overview of the presented topics

Image Processing, Graphics	Colour Perception	Additive/ Subtractive Colour	Graphic Formats	Printing	Colour Depth	
Coding	Morse Game	Creating a Code with Colours	Code Trees	Binary Numbers	Error Detection	Huffman-Code
Encryption	Caesar Cipher	Symmetric-key Cryptography	Public-Key Cryptography			
Hardware	Disassembling computers, Animations adequate for different ages					
Operating Systems	OS as a Shop	Deadlock Prevention	Scheduling			
Computer Networks	Chinese Whispers	Communication Rules	Postman-Game	Addressing and Protocols	Routing	
Algorithms	Instructions how to get somewhere	Structured Instructions				
Sorting	Selection Sort	Binary Search-tree	Merge Sort			
Searching	Blind Search	Searching in a linear Structure				
Automata Theory	Finite State Automata	Pushdown Automata				

In all lessons, the pupils cooperate in groups. Depending on the topic they act either as part of the computer, serving as data or as object being manipulated by algorithms, or assuming some role of a program.

Out of principle, computers were specifically not used during the lessons. The pupils learned, based on activities, simulations, and animations. Important didactical principals behind the concept are discovery learning and teamwork. The main goals of the project are:

- To increase/create pupils' interest in CS
- To especially motivate girls
- To broaden the pupils' attitude towards CS
- To show teachers that teaching proper CS topics is not as difficult as many of them assume.

The main focus of this paper is to show differences of the project's success in primary school and secondary school. Before describing the evaluation methods and results one module out of the topics in table 1 is presented as an example in the following chapter. All teaching units are available in German on the project web-page http://informatik-erleben.uni-klu.ac.at. In section 3 we describe the evaluation instruments and the obtained results.

2 Automata Theory in Primary Schools

Since not all topics taught in the classrooms can be shown in this paper, an example presenting some aspects of automata theory is introduced. The unit was taught in several classes.

In our daily life we find automata everywhere and in numerous variants such as coffee or chewing gum vending machines. The states and actions of these automata are comprehensible for children as well. Consequently, they are a good introduction into the field of theoretical informatics. In a 4th grade of a primary school automata theory was introduced by pupils "playing" or animating a chewing gum vending machine.

The unit (90 minutes) began with some questions like "Do you know some automata/machines? What can you do with these machines? What happens in a chewing gum vending machine? Which steps are necessary in order to get the product? What kinds of states are there (e.g. ready or finished)? What does the vending machine do? How does it know what to do? What happens when we don't insert enough money?"

Based on the questions and answers the main concept and vocabulary of finite-state machines were explained: There is always a start and a final or accept state, an input alphabet and one or more transition actions depending on the input and/or the actual state. All states and actions can be described by a transition diagram or a transition table which will both be shown and animated in this unit [3].

After these explanations the state diagram (figure 1) of a vending machine was presented by a poster that the children should learn to read.

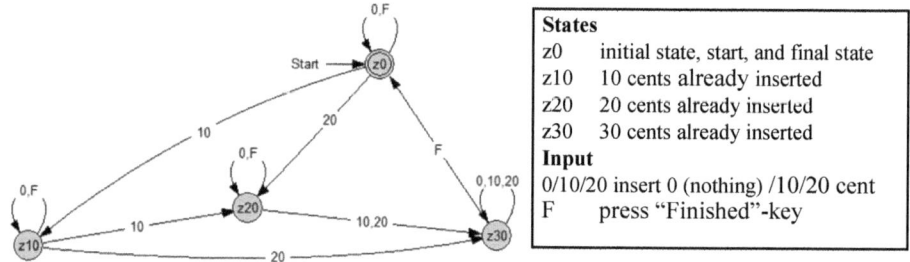

Fig. 1. State diagram of a chewing gum machine

The "route" from entering some money to receiving the chewing gum that costs 30 cent, was slowly shown on the diagram accompanied by questions and instructions like the following:

"The chewing gum vending machine is ready to start; it is in the initial state 'z0' (indicated by the circle at the top of the diagram. If we don't insert anything '0'[2] (0 Cent) or press the key 'F' (finished), the machine remains in the same state (indicated by the transition arrow labeled with '0, F'). What happens when we insert a 10 cent piece? – The machine passes to the state 'z10' (indicated by an arrow labeled with '10') and waits for more money. Which state will follow when we insert 20 cent more ("z30")? What happens, when we insert already 20 cent at the beginning? How much do we have to insert now?"

---

[2] Representing "insert nothing" by "0" has been used as didactical vehicle to transit from the operators, i.e., the pupils, view to the machine view. Thus, these superfluous zeros are explained away during the process.

Besides the diagram, the transition table for the same machine has been shown and read step by step as well. The children were then asked which type of presentation they preferred – the table or the diagram. Which one was easier to read and to understand? The answer was interesting: only four boys preferred the transition table. The other 15 pupils (5 boys and 10 girls) said that the diagram was easier to follow.

Then the pupils were asked to animate this vending machine. At first, the children representing the states (z0, z10, z20, z30) got a label and were placed at the front of the classroom according to the diagram on the poster. Those who didn't want to act should observe the animation and try to give instructions. The child representing the final state z30 got also a box with chewing gums. One child, who wanted to "buy" a chewing gum, got coins of 10 and 20 cent. He had to go to the child representing the start state z0 and give him a 10-cent-coin. The "start state" had to pass the coin to the "state" z10 who requested the buyer to give him additional 20 cents. The whole 30 cents were given to the child representing the final state z30. Then the buyer had to press the key "F" (for "Finished") and finally got a chewing gum. At last, the chewing gum machine is ready again, so the final state in this example corresponds to the initial state z0. In a second turn, another child could buy a chewing gum with another variation of coins.

After this animation the children should design their own machine. They were divided into four groups and each group chose another machine. Three groups designed automata corresponding to the presented vending machine for a pizza, a candy and a cup of coffee. One group of four boys tried to design a mobile telephone. This group was very engaged and interested. Certainly, the diagram wasn't perfect, but in view of the time available for this lesson the result was surprisingly good.

As in the described unit all other modules try to activate the pupils by playing animations.

After this demonstrative example, the paper presents the evaluation methods used and results of project.

3 Instruments and Results

Lessons on topics mentioned in table 1 were conducted in several classes and were evaluated by a spectrum of methods, including the teachers' opinions as well as the pupils' feedback. The following fields were investigated:

- Pupils' interest in informatics.
- Pupils' perception of informatics.
- Pupils' attitude towards the interventions.
- Teachers' willingness to teach the concept on their own.
- Differences between girls and boys, regarding the interest in informatics.

3.1 Population and Instruments

The subjects were pupils from primary school, lower secondary school and upper secondary school. Altogether 18 groups/classes participated in the first session of *Informatik erLeben* with a total number of more than 300 students. For the evaluation we tried different approaches. Each class participated in three lessons of *Informatik erLeben*, which took about two hours each. Table 2 shows the distribution of the participating classes with the grades and age of the pupils.

Table 2. Population and grades

Grade	Age of the pupils	Number of classes/groups
Primary School	6-10	7
Lower Secondary School	10-15	8
Upper Secondary School	15-19	3

To get an overall impression of the effects on pupils and teachers, five evaluation instruments were used. In upper secondary schools we used only a few evaluation instruments due to the low number of groups. The following chapters describe the methods of evaluation and the resulting outcomes.

3.2 Comparing Terms Concerning or Non-concerning CS

It was not easy to find an instrument to evaluate the views and attitudes of the students. We tried an approach with questions that were given to the pupils before and after our interventions.

Each pupil should answer two questions out of the following pool:

- What can you do with informatics?
- What do you consider to be the most interesting aspect of informatics?
- Where are you successful with informatics?
- Informatics is for me.

After the given interventions, the answers were compared and analyzed to determine if the answers were more related to concepts of informatics than before the interventions. Table 3 shows examples of statements mentioned by pupils.

Table 3. Categorization of terms for the evaluation (examples)

CS related	Not CS related
Learn a lot about technology	Computer games
Technology	Chatting
Bus	Writing
Calculating	Printing
Structure of a computer	To search information
Graphics file format	Listening to music
Processor	Watching videos
Main memory	Internet shopping

The comparison of the answers has shown obvious differences before and after the interventions.

Table 4. Percentage of pupils mentioning CS-related concepts

	Before the interventions	After the interventions
Primary school	16,9% (n=166)	49,4% (n=162)
Lower secondary school	18,8% (n=48)	50% (n=18)

As shown in table 4 about half of the pupils could be influenced in their view on computer science by the three interventions at school. Compared with the evaluation results described in [5] this is a relatively high percentage and shows that the evaluation instrument was adequate for this task. Still, pupils from lower secondary school had a more concrete idea of the subject than before our interventions.

3.3 Observation of Pupils' Attention

The observation of the pupils' attention involved the teachers observing their students. Based on the assumption that pupils are attentive only if they appreciate the situation, the results show whether a pupil likes the lesson or not. To assess the pupils' attention, teachers were asked to note each pupil's attention in a form containing the following scale (adapted from [4]):

- *On-task passive*: pupil follows the lesson passively.
- *On-task active:* pupil follows the lesson actively on his/her own.
- *On-task reactive:* pupil follows the lesson actively by reacting to a question.
- *Off-task passive:* pupil doesn't follow the lesson, but doesn't disturb.
- *Off-task disturbing:* pupil doesn't participate to the lesson and disturbs.

The form was divided into four categories according to the different teaching methods used during the lessons. With multiple entries, the teachers could express the pupils' attention. The units presented in the classes varied and, therefore, the total number of pupils observed in a particular category varies.

Intermediate results of the pupils' attention were already published in [2]. With the final evaluation of the first project-round all observation results are now available.

The observation of the pupils' attention has shown that the younger pupils were more attentive than the pupils from secondary school. The tables show a shift from the status of on-task active to the status of on-task passive with increasing age of the students.

Table 5 shows that most of the primary school kids were active during the lessons in all methodical categories. This attention shows that they liked the lessons and that they were interested in the topics.

Table 5. Attention of the primary school kids

	Lecture part (n=147)	Observation and animation (n=164)	Individual work (n=78)	Pair/group work (n=147)
On-task passive	15	12	6	14
On-task active	105	155	69	106
On-task reactive	40	45	29	27
Off-task passive	3	3	1	0
Off-task disturbing	2	0	0	1

Table 6. Attention of the lower secondary school pupils

	Lecture part (n=113)	Observation and animation (n=87)	Individual work (n=53)	Pair/group work (n=40)
On-task passive	25	22	11	1
On-task active	82	70	50	39
On-task reactive	8	1	0	0
Off-task passive	2	1	0	0
Off-task disturbing	2	0	0	0

Table 7. Attention of the upper secondary school students

	Lecture part (n=26)	Observation and animation (n=15)	Individual work (n=6)	Pair/group work (n=18)
On-task passive	11	8	2	5
On-task active	6	5	6	13
On-task reactive	3	2	0	0
Off-task passive	4	0	0	0
Off-task disturbing	2	0	0	0

The pupils from lower secondary schools followed the lessons as well but some of them were especially passive during the lecture part. Whereas most of them worked actively during the student centered parts. A continuation of this trend can be observed concerning attention of students from upper secondary schools.

Generally one can say that primary school pupils were more enthusiastic and active during the lessons compared to older pupils. However, the older pupils were also active, especially during the learner centered parts of the lessons. This indicates that also in other subjects learner centered methods should be preferred by the teachers.

3.4 Student Questionnaire

After finishing the project, a questionnaire was given to some classes. The aim of the questionnaire was to investigate the interests of the pupils and to determine, if they liked the lessons as well as to detect gender differences. The most important results are presented below. Table 8 shows the population that was given questionnaires.

Table 8. Population for the questionnaire

Grade	Number of pupils
Primary school	27
Secondary school (10-12 years old)	97
Secondary school (13-15 years old)	13

Describing the entire results and questions of the questionnaire used would exceed the scope of this paper. Only the most interesting results are described below. The entire results are described in [6].

Evaluation of the questionnaire has revealed that primary school kids are still much more interested in informatics than kids from lower secondary school. Fig. 2 shows the decline of interest with increasing age of the pupils.

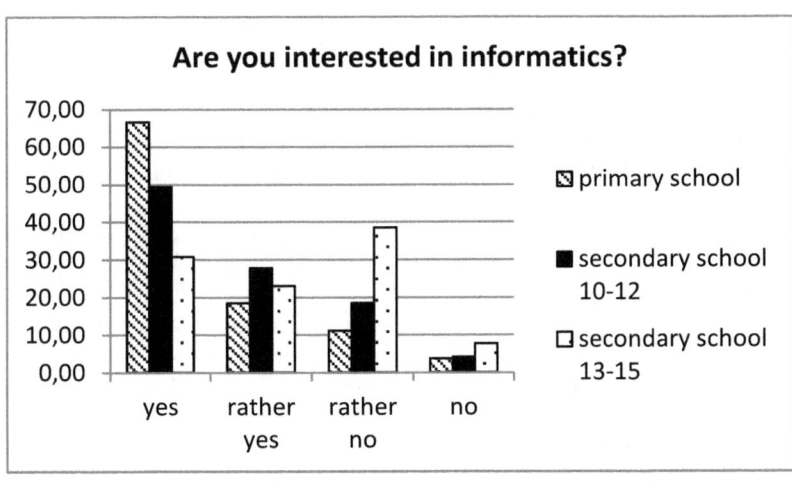

Fig. 2. Interest in informatics (n=27, 97, and 13)

In primary schools more girls (93%) are interested in informatics than boys (75%). In lower secondary schools this changes and in both age categories (10-12 and 13-14) more boys are interested. 100% of the primary school pupils think that after the interventions, the term informatics is clearer for them and that they understand it better.

In another question, the children were asked if our interventions at school changed their interest. Fig. 3 shows the interesting result, that the younger the pupils are, the more easily they can be influenced in their interest. Considering gender, especially girls from primary school could be influenced. All girls were interested in informatics after the interventions.

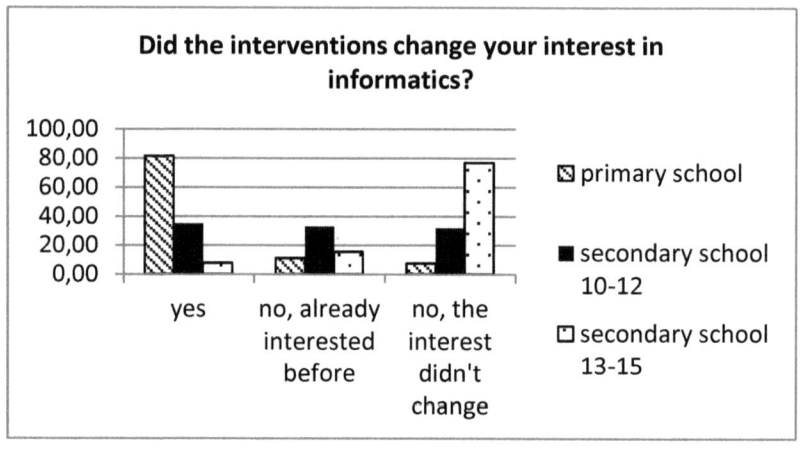

Fig. 3. Change of interest due to the interventions (n=27, 97, 13)

Reflecting on these results yields that it is very important to start at an early age to broaden the pupils' image of CS and to create interest. Especially in primary schools the interventions were very successful. While some boys were already interested in informatics before, all participating girls could be influenced. These findings are important for future work in encouraging girls for technical fields. It would not be enough to just create the interest in the primary school. Pupils must have the possibility to attend exciting CS lessons during all grades. Because primary school kids are very open and enthusiastic towards new topics and concepts, it is necessary to bring more technical topics in all primary schools.

Another question has revealed that gender-stereotypes are growing with the pupils' age. As shown in Fig. 4 most children from primary school think that girls are as talented for computer science as boys.

About 50% of the pupils from secondary school between 13 and 14 still think that boys and girls are equally suitable for computer science. Most of those who don't agree, think that boys are more suitable for computer science. Looking at the gender specific result, it can be said that especially boys think that they are more suitable for computer science than girls.

This gender specific shift underlines the fact that technical interest should be increased already at an early age.

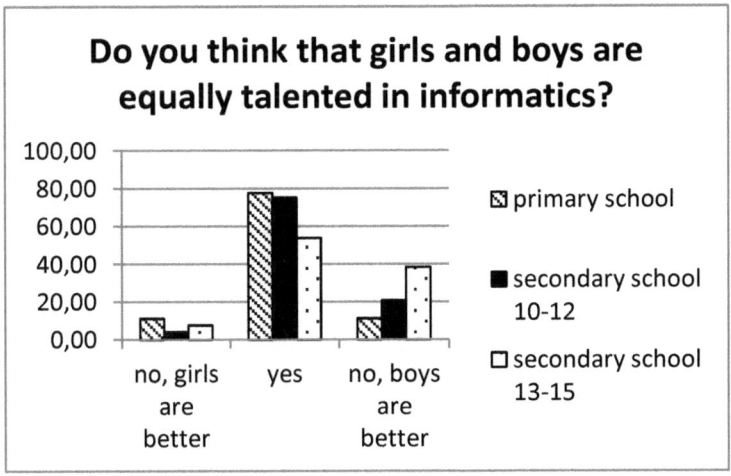

Fig. 4. Suitability for informatics (n=27, 97, 13)

3.5 Feedback Letters from Pupils

Some of the teachers asked their pupils to write feedback letters to the university teachers. These letters usually don't have a special format and gave feedback either in the whole project or on special lessons.

The voluntary feedback letters from the pupils support the findings from the interviews described so far.

Except for one class from a secondary school, the feedback of the pupils was totally positive. They appreciated the topics and the teaching methods. Especially girls, particularly those from secondary school, mentioned the alternative teaching methods that were learner centered. This feedback correlates with the findings regarding the attention described in 3.3.

3.6 Teachers' Interview

Teachers from all grades were interviewed. The table shows the number of interviewees for each grade.

Table 9. Number of teachers interviewed

Grade	Number of interviews
Primary school	6
Lower secondary school	4
Upper Secondary School	2

The goal of the interviews was, on the one hand, to find out the attitude of the teachers towards the concept of the interventions, and on the other hand, to see their opinion about the pupils' interest. To motivate the teachers to partake in the interviews we decided that the interviews should not be longer than about 10 or 15 minutes. This had the negative effect that the questions were rather superficial.

From the 18 groups taking part in the project 12 teachers were willing to do an interview. The twelve teachers interviewed confirmed the results about the pupils' interest. Only a few teachers considered the interest of their pupils lower than the pupils themselves.

The more important goal was to find out the attitude of the teachers towards teaching core-concepts of Informatics in the presented way, i.e., without computer and in a playful manner. All teachers appreciated the attempt to interest their pupils for computer science. In Austria there is no subject "Informatics" in primary school. All primary school teachers mentioned that there is a lack of technical topics in the curriculum of primary schools. They liked the interventions and are willing to teach the topics on their own. Some teachers from secondary school have doubts, though, that they are able to teach these topics.

It should be mentioned, that in Austria, most teachers who teach Informatics have not studied computer science. They originally studied other subjects and got their informatics education via in-service courses. Because of this, they all appreciated the didactical hints and the teaching material provided on the project page (http://informatik-erleben.uni-klu.ac.at).

4 Conclusion

The evaluation has shown that teaching Computer Science in primary school by the approach of *Informatik erLeben* was very successful. In fact the evaluation has revealed that especially pupils from primary school could be enthused with our lessons. The success declines with the age of the pupils/students. This underlines the importance of such initiatives already at an early age. With the declining number of CS students, also at the international level, as reported in [7], such initiatives are becoming important in many countries. As the example in this paper shows, core-topics of Informatics can easily be taught in primary school at an adequate level.

We hope that our positive experiences with young children will motivate others to teach more technical topics at primary school.

References

1. Bell, T., Witten, I.H., Fellows, M.: Computer Science Unplugged. An enrichment and extension programme for primary-aged children, http://csunplugged.org/ (April 5, 2011)
2. Mittermeir, R.T., Bischof, E., Hodnigg, K.: Showing Core-Concepts of Informatics to Kids and Their Teachers. In: Hromkovič, J., Královič, R., Vahrenhold, J. (eds.) ISSEP 2010. LNCS, vol. 5941, pp. 143–154. Springer, Heidelberg (2010)
3. Hopcroft, J.E., Ullman, J.D.: Introduction to Automata Theory, Languages, and Computation. Addison-Wesley, USA (1979)
4. Helmke, A., Renkl, A.: Das Münchner Aufmerksamkeitsinventar (MAI): Ein Instrument zur systematischen Beobachtung der Schüleraufmerksamkeit im Unterricht. Diagnostica 38(2), 130–141 (1992)

5. Taub, R., Ben-Ari, M., Armoni, M.: The Effect of CS Unplugged on Middle-School Students' Views of CS. SIGCSE Bulletin 41(3), 99–103 (2009); Proc. 14th SIGCSE on Innovation and Technology in CS Education (ITiCSE 2009)
6. Bischof, E.: Interventionen im (Informatik-) Unterricht. Ein Versuch bei SchülerInnen das Bild der Informatik zu erweitern. Dissertation. Alpen-Adria Universität Klagenfurt (2011)
7. Curzon, P., Cutts, Q.I., Bell, T.: Enthusing & Inspiring with Reusable Kinaestetic Activities. SIGCSE Bulletin 41(3), 94–98 (2009); Proc. 14th SIGCSE on Innovation and Technology in CS Education (ITiCSE 2009)

Pre-service Computer Science Teacher Training within the Professional Development School (PDS) Collaboration Framework

Noa Ragonis and Anat Oster-Levinz

School of Education,
Beit Berl College,
Doar Beit Berl, 44905, Israel
{noarag,anato}@beitberl.ac.il

Abstract. The professional development school (PDS) represents a collaborative framework between high schools and teacher training institutions. The main objective of the PDS is to deepen the student teachers' involvement in the school, to expose them to the variety of teacher work tasks, and to provide them with a wide range of experiences. This paper focuses on the development and implementation of a computer science (CS) student teacher practicum program that takes place within the PDS framework. We present the significant advantages that can be gained using PDS principles and demonstrate different activities performed in order to gain them within the training of CS student teachers. The paper reflects the accumulative knowledge acquired over the past five year at the Beit Berl College, Israel. We believe that this paper will contribute to the community of CS teacher preparation educators.

Keywords: CS student teachers, CS practicum, CS teachers' preparation program, CS PDS practicum, PDS.

1 Introduction

A great deal of attention has been devoted in recent years to the importance of computer science (CS) teacher preparation programs. Such training programs are recognized as one of the key components in the establishment of successful high school CS curricula [10]. This increased attention is reflected in the number of publications devoted to CS teachers training [1, 2, 9, 15, 16, 18] and in several publications of the Computer Science Teachers Association (CSTA), the leading CS teachers organization in the US, which focus on CS teachers training [4, 5, 21].

The two main teachers training tracks offered are: (1) Teaching certificate studies offered to graduate students who already have a bachelor's degree in CS; (2) Bachelor of Education (B.Ed.) studies that include disciplinary studies alongside extensive educational studies and also provide a teaching certificate. Both tracks include three main tiers: disciplinary studies, general education and pedagogical studies (e.g., educational psychology), and pedagogical-disciplinary studies, which focus on

I. Kalaš and R.T. Mittermeir (Eds.): ISSEP 2011, LNCS 7013, pp. 106–116, 2011.
© Springer-Verlag Berlin Heidelberg 2011

principles and tools used to teach concepts included in the CS knowledge domain. The pedagogical-disciplinary studies relate to the pedagogical content knowledge (PCK), which refers to what a teacher is required to know in order to teach a certain subject matter, how to make it understandable, learners' preconceptions and misconceptions, and strategies for coping with learners' misconceptions [19], [20]. PCK is usually acquired in a Methods of Teaching CS course and during a practicum in school classes.

This paper focuses on the practicum of CS student teachers in school classes and presents an example of a practicum structure that implements the professional development school (PDS) framework for training CS student teachers. As background information we will present the common kind of practicum that exists in teacher preparation programs, followed by a presentation of the main principles of the PDS framework and how those principles are reflected in the extensive practicum. Finally, we will present the setting and the method of implementing the PDS practicum among CS student teachers at the Beit Berl College. The potential contribution and benefits that can be gained from integrating the PDS practicum model will be demonstrated using specific examples.

2   Practicum Background

Most teacher preparation programs include a practicum in school classes. The main objectives of the practicum are to enable the student teachers to experience their future field of work and to comprehend the scope and uniqueness of real teaching before becoming CS teachers. Here we address the more common kind of practicum and discuss the rationale and implementation of the comprehensive practicum in the spirit of the PDS.

2.1   The Practicum in Teacher Preparation Programs

The in-school practicum is usually based on two activities performed by the student teachers, namely observing lessons and teaching lessons. The extent of the practicum activities varies among institutions. Some programs require that the practicum be performed for a specific, relatively short period; others require a full year of participation in school activities [8]. One example of limited practicum requirements is to observe five lessons and to teach three lessons. Hazzan and Lapidot [11] address the attention given to the central rule of the student teachers' practicum, by recommending ways in which it can help bridge gaps between theoretical knowledge and actual performance. Another framework for bridging this gap is a tutoring program, which is integrated into teacher preparation programs, in which student teachers guide learners in problem-solving processes, as presented in [17].

2.2   The PDS Practicum Framework

The difficulties experienced by new teachers entering their field of work are widely recognized as a major cause of the high dropout rate of new teachers from the school system. The PDS collaboration framework aims to assist teachers in taking their first steps in their professional career.

The rationale of the professional development schools is to form innovative foundations through partnerships between schools and professional education programs within academic institutions [3]. A PDS is a collaborative effort to improve the initial preparation of teachers and to enhance the professional development of classroom teacher. A PDS is a learning-centered community and its partners are guided by a common vision of teaching and learning that is grounded in research and practitioner knowledge. According to the PDS approach, the collaboration enhances change and mutual development in both systems (academic institute and school) and minimized the gap between them [6, 13, 14]. This empowers the prospective teachers and gives them a broader base for their future work as teachers. Through the PDS framework, student teachers can gain a unique benefit from a real teaching experience and form strong connections between theory and practice.

The implementation of the PDS is strongly based on collaboration whose objectives within a specific school are determined jointly by all partners: school administrations, subject-matter teachers, and the teacher preparation institute. The integrated objectives are designed to advance the mutual interests of all partners. The building of a learning community is based on the sharing of disciplinary and pe-dagogical knowledge, reflective processes, research on the teaching of the discipline, bridging between theory and practice, construction of teaching-learning activities, planning of long-term projects, and encouraging initiatives [22].

CS student teachers at Beit Berl College experience this extent and collaborative type of practicum, as presented and illustrated in this paper.

3   The PDS CS Practicum Setting

3.1   The Student Teachers

Student teachers who participate in the practicum at Beit Berl College are studying towards a high school CS teaching certificate, which is mandatory in order to become a teacher in Israel. The student teachers come from two main populations: (1) student teachers studying for a B.Ed. degree, which includes CS disciplinary studies; (2) student teachers who already have a BSc degree in CS and wish to obtain a teaching certificate as well. Approximately 8-15 student teachers participate in the practicum each year.

3.2   The High School CS Curriculum

CS is an elective subject-matter in the Israeli high school curriculum. The Israeli high school CS curriculum comprises five units (90 school-hours each), as presented in Table 1, which are covered by the student teachers' practicum in the high school [7].

Table 1. The Israeli high school curriculum

Unit	Content
Units 1-2	Fundamentals of computer science (in Java or C#)
Unit 3	Choice of an alternative programming paradigm: logic programming, functional programming, or internet programming.
Unit 4	Software design, modularization, and ADT: stacks, lists, and trees
Unit 5	Choice of a theoretical topic unit: automata theory, object-oriented programming, or operations research.

3.3 Main Objectives of the Practicum

The objectives of the student teachers' practicum within the PDS in a specific school are defined on three levels [12]:

(1) Social-institutional collaboration – exposing the student teachers to the school as an active institution and involving the students in the school activities. Examples of such activities are: facilitating a volunteer project, coping with challenges of special-needs pupils, and collaborating with CS teachers in routine tasks during recesses throughout the school day.

(2) Collaboration in the teaching field – actively involving the students in various teaching activities, such as establishing and running online learning days programs at the school, guiding research and projects, providing weaker students with individual assistance, integrating student teachers into teaching and into writing, assessing, and marking of examinations.

(3) Professional advancement – developing a professional learning community in the school that comprises student teachers, pedagogical supervisors, and mentor teachers. We consider this level to be the most important one.

4 Training CS Student Teachers within the PDS

In this section, we present the training process of CS student teachers at Beit Berl College within the PDS collaboration framework. The implementation of the PDS is unique to the CS practicum and was established by the second author of this paper. In what follows, we address the schedule of a typical day of practicum, the method of implementing the practice of teaching, the rationale and implementation of a full week of practicum, some additional practice activities, and the evaluation of CS student teachers within the PDS collaboration framework. In our discussion we use the following definitions: a *student teacher* is a student who is enrolled in the academic practicum course, a *mentor teacher* is a subject-matter high school teacher involved in the practicum of the student teachers, a *pedagogical supervisor* is a lecturer at an academic institute who teaches the practicum course and is present during all of the student teacher's practicum days at school, and a *pupil* is a high school student.

4.1 The Schedule of a Typical Day of Practicum

Each student teacher practices at school for one day a week throughout the entire academic year. At the beginning of the academic year, a meeting is held between the student teacher, the mentor teacher, and the pedagogical supervisor with the objective of constructing a schedule for the student teacher's practicum day at school. Each student teacher has a daily schedule that is both individual and flexible and can be changed during the school year to include various experiences.

A typical practicum day at school consists of a combination of the following activities:

(a) Observing lessons - The student teachers observe lessons in the mentor teacher's classes, including lab lessons.

(b) Teaching lessons - The student teachers teach lessons, gradually expanding their extent. At the beginning of the year, they teach only a part of a lesson; for example, they may explain part of a program to the class. Later on, they may teach a full problem-solving process progressing to teaching an entire lesson, for example, on nested If statements. By the end of the year they teach an entire teaching unit, for example, arrays.

(c) Active participation - The students teachers are active participants in CS staff meetings (the learning community), in which various teaching issues are addressed, for example, the issue of integrating visualizations.

(d) Developing and grading tests - The student teachers develop tests (for example, a final test on loops) together with the mentor teacher and the pedagogical supervisor, and help the mentor teacher grade the pupils' tests.

(e) Supporting pupils with special needs - The student teachers help pupils with special needs during tests, for example by reading and dictating tests.

(f) Fulfilling teacher obligations - Student teachers are also involved in other daily obligations of teachers, for example, monitoring pupils in the school yard during recess, a duty all teachers must fulfill.

(g) Personal meetings - Student teachers hold personal meetings with the mentor teacher and pedagogical supervisor.

(h) Group discussions - The student teachers, as a group, have discussions with other student teachers from the same discipline or from other disciplines on various class learning abilities, successful events, and difficulties that arose while teaching.

(i) Tutoring pupils - Student teachers tutor pupils in their problem-solving processes, guide them and help them overcome their obstacles.

Since it is clear that all of the above activities cannot be achieved during each and every day of the practicum, the student teachers' schedules are flexible so that they have the opportunity to practice the various activities throughout the entire year. Table 2 presents an example of a student teacher's practicum schedule at the beginning of the year.

Table 2. Example of a student teacher's day of practicum

Lesson	Program
1	Observing a Grade 12 computer lab lesson given by the mentor teacher.
2	Observing a Grade 12 lesson on binary trees given by the mentor teacher.
3	Meeting with the mentor teacher and pedagogical supervisor concerning issues that arose during the binary trees lesson.
4	Computer science staff meeting. Main issue: composing a Grade 10 test on If statements.
5	Meeting with the pedagogical supervisor in preparation for teaching a lesson on two-dimensional arrays in Grade 11 that will take place the following week.
6	Meeting with the pedagogical supervisor and all CS student teachers in preparation for the tutoring activity.
7	Tutoring Grade 10 pupil. Main topic: If statements.

In addition to the various activities discussed above, the student teachers are also responsible for leading special initiatives in the school. Such initiatives are usually not directly related to CS, but school officials consider the CS students teachers and their pedagogical supervisor as part of the professional information and communication technologies (ICT) community. For example, student teachers may lead an Online Day in which all students stay at home and participate in online activities in various disciplines. The CS student teachers help the school teachers develop the online activities and help them activate the activities throughout the Online Day. Another example of an ICT school activity is running a simulation program for computerized elections, which student teachers can develop during a national election year and which the school can use on Election Day to simulate an election process.

4.2 Practice Teaching

Preparing a lesson to be taught by a student teacher requires a cooperative learning process in which the student teacher consults and collaborates with the pedagogical supervisor and mentor teacher.

The CS student teacher must:

(a) coordinate the lesson time and topic with the mentor teacher;
(b) establish aims and operative goals for the lesson;
(c) plan the lesson sequence in consultation with the mentor teacher and/or pedagogical supervisor (at the beginning of the year the student teacher receives detailed guidance for lesson planning and by the end of the year he or she is expected to act independently);
(d) plan tasks and activities related to the aims and operative goals that are compatible with the learners' knowledge, abilities, and diversity. The lesson can incorporate various kinds of classroom activities (e.g. whole-class,

pair/group work activity, lab activity) and student teachers should use a variety of teaching tools (e.g. media, games, role playing, and simulations);

(e) develop all required tasks and activities in detail, including solutions, and allocate the time required for each task or activity (i.e. make a lesson timetable). The student teacher should be aware of the logical progression of the sequence of activities;

(f) discuss the lesson plan with the pedagogical supervisor and mentor teacher (online or in person). Such conversations may deal with the best solution to be presented to pupils in class or they may focus on the order of the tasks or on how to explain the task or the solution.

Ultimately, the student teacher is responsible for all of the above.

In the sequel, we present an example of a discussion between a student teacher (ST) and his pedagogical supervisor (PS). The two had previously decided that the student would teach in a 12-grade class a lesson on how to insert a new element into a sorted list.

Part of the online conversation between the student teacher and the pedagogic supervisor regarding the sequence of the lesson was as follows:

ST: I'm wondering how to start the lesson.
PS: There are several ways to start a lesson; we spoke about them last week…
ST: I'm thinking about demonstrating a chain on the board….
PS: What about a role play with the students?
ST: I don't think that will work with 12-graders.
PS: You'd be surprised to see the pupils' enthusiasm about participating in a role play!

Ultimately, the student teacher initiated a role play in the class: the pupils' role was to be elements in a sorted list and one pupil was the new element that was to be inserted into the proper position into the sorted list. The demonstration focused on several different cases: (1) The element must be inserted as the first element in the list; (2) the element must be inserted at the end of the list; and (3) the element must be inserted into the middle of the list.

When a student teacher teaches a lesson, all other CS student teachers in the course are invited to observe it. A feedback discussion is conducted at the end of each student teacher's lesson, led by the pedagogical supervisor. The mentor teacher and the other CS student teachers, who observed the lesson, participate in this discussion. The feedback session starts with the student teacher who addresses his or her feelings during the lesson, reflects on the lesson plan compared with its implementation, and on what in his or her opinion was done properly and what needs to be improved. The student teacher directs the discussion to the issues he or she deems important.

CS student teachers usually experience more teaching than do student teachers in other disciplines. This is because the practicum must cover both regular classroom lessons and computers lab lessons, which provide the CS student teachers with many opportunities to teach small groups of pupils. The pupils in 10th and 11th grades are highly heterogenic and diverse, so mentor teachers use the opportunity to divide the class into smaller and more homogenous groups, thus providing a more suitable learning process for the pupils and a good practice opportunity for the student teachers.

4.3 A Full Week of Practicum

In addition to the individual practicum days, the annual practicum plan includes two full weeks of practice at school, one in each of the two semesters, during which there are no lectures at the college and the student teachers spend the entire week at school. This is a unique opportunity to experience the continuity of the teacher's work.

The goals of the week-long practicum (as opposed to single days) are: (a) to expand the student teachers' opportunities for acquaintance with the school system, including all of its strata and activity settings, which student teachers are unable to experience during their one-day-a-week practicum; (b) to enable the student teachers to teach a sequence of lessons; (c) to give student teachers an opportunity to authentically experience the teacher's work during an entire week, which helps them understand the complexity of the teacher's work at school; (d) to enable a meaningful dialogue with mentor teachers, school staff, other student teachers, and the pedagogical supervisor in order to fully understand the underlying aspects of various educational processes; and (e) to enhance informal relationships among student teachers, mentor teachers, and high school pupils, mostly as practice for developing relationship with their future teaching partners.

It is our experience that a full week of practicum enriches all PDS partners. During this week, the CS student teachers teach in different classes, something they never experience during a single day of practicum, enabling them to become acquainted with various learning methods. They also have the opportunity to sit in on homeroom classes and participate in other school activities.

During the two weeks of practicum, the student teachers also join their pedagogical supervisors on visits to various high-tech companies and other high schools that implement different alternatives of the CS curricula or have special, state-of-the-art computer labs. The objective of these activities is to expand as much as possible the student teachers' school experience, as well as their CS knowledge.

4.4 Evaluation of the Student Teachers

The evaluation of student teachers within the PDS collaboration framework refers to the entire process that they undergo. The student teachers' teaching skills are evaluated as is their performance in all other activities that take place within the PDS collaboration partnership. The main objective of the evaluation is to monitor the student teacher's improvement throughout the process, from the beginning of the school year to its end. Table 3 presents the main components of student teacher evaluation within the PDS collaboration framework.

Table 3. Student teacher evaluation components within the PDS collaboration framework

Content Knowledge

Demonstrates ability for self-education.

Demonstrates knowledge of the relevant contents of instruction required for teaching CS in school.

Provides multiple and varied explanations, examples, and details to support the topic.

The content of instruction suits the students' level.

Pedagogical Knowledge

Structure of the lesson: Detailed lesson plan; opening of lesson; logical progression/sequence of activities; smooth transitions from stage to stage; timing, pacing of the lesson; closure.

Teaching Performance: Demonstrates knowledge of the school's CS curriculum; demonstrates effective use of questioning skills; exhibits strong presence in class; provides effective feedback; gives clear instructions; uses the board and/or other instruction media effectively; uses program visualizations in class; monitors students' work; uses effective management techniques (discipline); incorporates various types of activities; uses materials and tasks that are compatible with the students' heterogeneity; provides opportunities for students to engage and pursue interests.

Classroom Environment

Creates a supportive learning environment (allows students to err and take risks, provides opportunities for every student).

Provides opportunities for peer/group interaction (mostly in the lab).

Professional and Collaborative Responsibilities

Is punctual; assumes responsibility; is polite; maintains school formalities; exhibits good rapport with students/peers/teachers; shows interest in the school system.

Is open to suggestions and comments.

Works well with peers and colleagues.

Shows initiative and is active in teaching as well as in other school activities (such as tutoring or Online Day).

Test lesson
The student teacher is graded on his/her teaching of a lesson.

5 Summary

The purpose of the traditional practicum is to develop the pre-service teacher's understanding of complex classroom situations associated with the teaching of the subject matter. The PDS collaboration framework promotes this understanding in a broader and deeper manner, and offers a comprehensive training program in which student teachers experience a variety of teacher tasks that relate both to the entire school structure and to the specific discipline taught. In this paper, we discussed the objectives and principles of the PDS, describing the unique training process for CS student teachers and highlighting the advantages of various activities that CS student teachers experience within this framework.

We will close with a quote from the practicum syllabi:

> *"Life at school is rich in events. The school and college staff invests great efforts in organizing a plentiful learning environment for you. All you must do is cooperate, take responsibility for your learning, and exhibit high motivation. Both staffs are more than willing to assist you."*

References

1. Bell, T., Lambert, L.: Teaching Computer Science Majors about Teaching Computer Science. In: 42th SIGCSE Technical Symposium on Computer Science Education, pp. 541–546. ACM, New York (2011)
2. Blum, L., Cortina, T.J.: CS4HS: an Outreach Program for High School CS Teachers. ACM SIGCSE Bulletin 39(1), 19–23 (2007)
3. Clark, R.W.: Effective Professional Development Schools: Agenda for Education in a Democracy, pp. 3–4. Jossey-Bass, San Francisco (1999)
4. CSTA: Computer Science State Certification Requirements - CSTA Certification Committee Report (2007),
   http://www.csta.acm.org/ComputerScienceTeacherCertification/
   sub/TeachCertRept07New.pdf (April 2011)
5. CSTA: Ensuring Exemplary Teaching in an Essential Discipline: Addressing the Crisis in Computer Science Teacher Certification, Final Report of the CSTA Teacher Certification Task Force (2008),
   http://www.csta.acm.org/Communications/sub/DocsPresentationF
   iles/CertificationFinal.pdf (April 2011)
6. Darling-Hammond, L.: When Conceptions Collide: Constructing a Community of Inquiry for Teacher Education in British Columbia. Journal of Education for Teaching 27(1), 7–21 (2001)
7. Gal-Ezer, J., Harel, D.: Curriculum and Course Syllabi for a High-School CS Program. Computer Science Education 9(2), 114–147 (1999)
8. Gal-Ezer, J., Hazzan, O., Ragonis, N.: Preparation of High School Computer Science Teachers: The Israeli Perspective. In: 40th SIGCSE Technical Symposium on Computer Science Education, pp. 269–270. ACM, New York (2009)
9. Grugurina, N.: Computer Science Teacher Training at the University of Groningen. In: Mittermeir, R.T., Syslo, M.M. (eds.) ISSEP 2008. LNCS, vol. 5090, pp. 272–281. Springer, Heidelberg (2008)

10. Hazzan, O., Gal-Ezer, J., Blum, L.: A Model for High School Computer Science Education: The Four Key Elements that Make it! In: 39th SIGCSE Technical Symposium on Computer Science Education, pp. 281–285. ACM, New York (2008)
11. Hazzan, O., Lapidot, T.: The Practicum in Computer Science Education: Bridging Gaps Between Theoretical Knowledge and Actual Performance. In: 35th SIGCSE Technical Symposium on Computer Science Education, pp. 29–34. ACM, New York (2004)
12. Klieger, A., Oster-Levinz, A.: In Search of the Essence of a Good School: School Characteristics Leading to Successful PDS Collaboration. Australian Journal of Teacher Education 33(4), 40–54 (2008)
13. Korthagen, F.A., Kessels, J.P.M.: Linking Theory and Practice: Changing the Pedagogy of Teacher Education. Educational Researcher 28(4), 4–17 (1999)
14. Levine, M.: Foreword. In: Teitel, L. (ed.) The Professional Development Schools Handbook: Starting, Sustaining and Assessing Partnerships that improve Student Learning, pp. XIII–XVII. Corwin Press, Inc., Thousand Oaks (2003)
15. Ragonis, N., Hazzan, O., Gal-Ezer, J.: A Study on Attitudes and Emphases in Computer Science Teacher Preparation. In: 42th SIGCSE Technical Symposium on Computer Science Education, pp. 559–564. ACM, New York (2011)
16. Ragonis, N., Hazzan, O., Gal-Ezer, J.: A Survey of Computer Science Teacher Preparation Programs in Israel Tells Us: Computer Science Deserves a Designated High School Teacher Preparation! In: 41th SIGCSE Technical Symposium on Computer Science Education, pp. 401–405. ACM, New York (2010)
17. Ragonis, N., Hazzan, O.: Integrating a Tutoring Model into the Training of Prospective Computer Science Teachers. Journal of Computers in Mathematics and Science Teaching 28(3), 309–339 (2009)
18. Ragonis, N., Hazzan, O.: Disciplinary-Pedagogical Teacher Preparation for Pre-Service Computer Science Teachers: Rational and Implementation, Informatics in Secondary Schools - Evolution and Perspective. In: Mittermeir, R.T., Sysło, M.M. (eds.) ISSEP 2008. LNCS, vol. 5090, pp. 253–264. Springer, Heidelberg (2008)
19. Shulman, L.S.: Those Who Understand: Knowledge Growth in Teaching. Educational Teacher 15(2), 4–14 (1986)
20. Shulman, L.S.: Reconnecting Foundations to the Substance of Teacher Education. Teach. Coll. Record 91(3), 300–310 (1990)
21. Stephenson, C., Gal-Ezer, J., Haberman, B., Verno, A.: The New Educational Imperative: Improving High School Computer Science Education, Final Report of the CSTA, Curriculum Improvement Task Force (2005) http://csta.acm.org/Communications/sub/DocsPresentationFiles/White_Paper07_06.pdf (April 2011)
22. Teitel, L.: The Professional Development Schools Handbook: Starting, Sustaining and Partnerships that Improve Student Learning, pp. 1–7. Corwin Press, Inc., Thousand Oaks (2003)

Teaching Theoretical Informatics to Secondary School Informatics Teachers

Daniela Bezáková and Michal Winczer

Department of Informatics Education,
Faculty of Mathematics, Physics and Informatics,
Comenius University, Mlynská Dolina, 842 48 Bratislava, Slovakia
{bezakova,winczer}@fmph.uniba.sk

Abstract. This paper describes an exploration and discovery-based introduction of theoretical informatics for (a) pre-service teachers of informatics and (b) in-service teachers of informatics who lack the qualification. It was taught at Comenius University in Bratislava, Slovakia. It shows how this particular approach allowed us to teach fundamental concepts of theoretical informatics in an engaging yet scientifically sound manner to students with a limited mathematical background. A short participant evaluation of this approach, based on questionnaires and interviews, is presented.

Keywords: pre-service and in-service teachers of informatics, introduction to theoretical informatics.

1 Introduction

This paper considers a long-standing problem: how to teach an introduction to theoretical informatics (TI) to secondary school pre-service and in-service teachers of informatics who lack sufficient mathematical background?

This is an important issue in Slovakia because students are entering universities with a weaker mathematical background than in the past. There is less mathematics taught at the secondary school level; several topics taught a decade ago are now absent from the compulsory teaching plans, and students are not fluent in the language of mathematics or in mathematical thinking. Additionally there are many in-service informatics teachers without sufficient mathematical background and with no informatics education. It is our belief that informatics teachers need to understand the principles of TI, at least at the introductory level. Without this understanding, informatics education tends to degrade to the straightforward use of ICT.

The following is a detailed description of the present situation of teacher education in Slovakia and our approach to teaching TI to that audience.

2 Present Situation

The situation described here is based on long experience with teaching pre-service and in-service secondary school teachers at Comenius University, Bratislava,

I. Kalaš and R.T. Mittermeir (Eds.): ISSEP 2011, LNCS 7013, pp. 117–128, 2011.

Slovakia. Secondary school teachers in Slovakia are required to have a specialization in two subjects (e.g. Slovak Language and History, Mathematics and Physics, Biology and Geography, etc.). They are allowed to choose virtually any combination of the two subjects. Their study consists of three equally important parts: one part is general pedagogy and psychology; the other two parts consist of the chosen subject areas combined with those subjects' specialized pedagogies.

The secondary school subject concentrated on here is called informatics. It deals with all aspects of digital technologies, focusing on basic principles rather than how to use the technologies and the necessary skills needed to use them.

In this article we focused on courses for informatics teachers covering theoretical informatics (Computer science). We worked with two groups of students:

- Pre-service teachers - students of the Faculty of Mathematics, Physics and Informatics at Comenius University. These students study the teaching of informatics, in combination with mathematics, physics, biology, geography or chemistry.
- In-service teachers of informatics who have no qualification for teaching informatics, but who are willing to obtain it.

The second category consisted of 200 teachers who participated in the Slovak national project, *Education of In-service secondary school teachers of informatics* (*DVUI* project 2009-2011). The goal was to target curriculum for these teachers who are already teaching informatics, but who have no formal preparation in the subject. The teachers in this group had been attending courses for five semesters. The course of study consisted of 57 modules structured in four groups: digital literacy, the modern school, specialized informatics content and didactics of informatics. Each module consisted of eight 60-minute periods. The specialized informatics content contained modules on programming, algorithms and data structures and TI [4, 5, 6].

2.1 More Detailed Description of the Student Groups

The following is a more detailed description of the expectations and mathematical knowledge of both groups prior to attending our courses in TI, in the hope that the reasons for this approach will be clarified.

Pre-service Teachers. As mentioned above, the pre-service secondary school teachers are prepared to teach two subjects. Future teachers can combine informatics with geography, chemistry, biology, mathematics or physics, but theoretically, the second subject might be arbitrary. The number of students in these combinations is generally small. It cannot be assumed that the incoming students have taken the comprehensive final exam *(abitur, matura)* in mathematics at the completion of their secondary school studies. According to our surveys 40% of the pre-service teachers at our own faculty have not taken this exam in mathematics, 73% of them have not taken this exam in informatics and 30% of them have not taken this exam in either mathematics or informatics. Future teachers combining informatics and mathematics have typically taken the exam in mathematics, in contrast to students with the combination of geography, chemistry, or biology.

Students combining informatics with a subject other than mathematics have substantially fewer mathematics lessons, and consequently their mathematical background is insufficient for studying the theoretical foundation of informatics at the usual university level. Therefore, the content of TI courses needs to be adjusted, to make it manageable, even for students with a weaker mathematical background.

All first-year students who intend to be future teachers of informatics attend a semester seminar (two hours per week) which has a primarily didactical goal *"to build concrete tangible foundations, to increase motivation and preconditions for a more successful encounter with elements of theoretical informatics"* [1]. The students explore prepared microworlds and become familiar with some basic concepts of TI. For example, they connect basic logical gates to implement a given logical function, they compose rules for generating L-systems and discover how the outcome can be visualized, or they play with ciphers, coding and encoding messages. In time, the content of the seminar naturally evolved and we added much more material in the spirit of the text by Hromkovič [2], including limits of computability, complexity theory, the concept of randomness and its importance in TI, DNA computing and quantum computers. Later in their study, students encounter some of the themes from this seminar in a more formal way in a course similar to the one described in part 3.1 of this article.

In-service Teachers. In the *DVUI* project, there were 200 in-service secondary school teachers of informatics without an informatics qualification. We taught a group of 34 such in-service teachers at our faculty. A majority of them, 28, were lower secondary school (middle school) teachers. Only six members of the group were upper secondary school (high school) teachers (four from vocational schools). Twenty of them were teachers of mathematics and nine of physics, while six were both math and physics teachers. The majority of the group was only familiar with very simple mathematics -- corresponding to lower secondary school or upper vocational school. A great number of them were teachers with a humanities background (history, arts and languages) who were educated in mathematics a long time ago and only at the secondary school level.

During the project, the participants completed three modules of mathematics, each consisting of eight hours of instruction time. It seems clear that one could not cover as much mathematics as necessary in such a short time.

Both groups of students considered in this study lacked the mathematical background that is typically required for courses in TI. A question arises immediately: should TI ever be taught under such conditions? The answer is YES! A way must be found to make TI accessible, to avoid formalism as much as possible, but without compromising exactness. In a way, these are contradictory requirements. In spite of this, we try to balance informality and exactness in the courses. One way to do this is to find clever examples or problems that enable the learners to gain the experience of discovering ideas and concepts for themselves.

3 Implementation

In order to implement our vision we try to follow a few useful principles:

- Teaching less may mean learning more.
- Concentrate on very basic and important concepts.
- As much as possible, use carefully chosen examples and let the students discover the relevant concepts.

At the same time we focus on two goals:

- to present the necessary information to the students to build knowledge and
- to do it in a didactical way that motivates the students and provides inspiration for their teaching practice.

It is important to make sure that the students fully understand the presented concepts, why they were introduced and how they are connected to practical issues. It is very important that teachers are ready and able to explore new challenges, possibilities and approaches together with their students. Teachers must be prepared to give hints and to clarify misunderstandings, to be more like partners and advisors than reproducers of information. Teachers should inspire their students and at the same time, students should inspire the teacher's work.

In the following section, an overview of the concepts introduced to the students in our course for secondary school teachers of informatics, and some examples of our approach to teaching these concepts are presented in more detail.

3.1 Overview of Taught Material

The *DVUI* project allocated three modules to TI, with a printed textbook created for each of them [4, 5, 6]. The first module dealt with the design of algorithms [4], the second dealt with more theoretical topics such as the finite automata and the Turing machines [5] and the third emphasized contemporary topics such as cryptography and randomized algorithms [6]. Our approach to the presentation of the concepts and ideas was similar to those used in [7, 8, 9, 10, 11]. However, the works mentioned either focused on different areas of TI or they have different target groups.

As an example, we describe how we taught the second module. We wanted to introduce the following concepts: *the simplest computational model – finite automaton, alphabet, word, language, configuration, computation step, computation, accepting and rejecting computation, language recognized by a finite automaton, simulation, non-existence of a finite automaton for a language, determinism, non-determinism, various possibilities of extending the finite automaton model and the Turing machine.* This list is quite long. The individual concepts are important but at the same time are abstract and difficult to understand at first encounter, especially if the student is not trained in formal mathematics. That is why we encourage the participants to play, and through that play, guide them to discover the concepts for themselves. Each new concept is connected to real situations.

Teachers started by solving five tasks from the Bebras contest [3], a highly popular event among secondary school students. Surprisingly, it was quite interesting for the teachers to solve tasks intended for their students. The selected tasks had elements in common with the diagrams shown in Fig. 1 and Fig. 2.

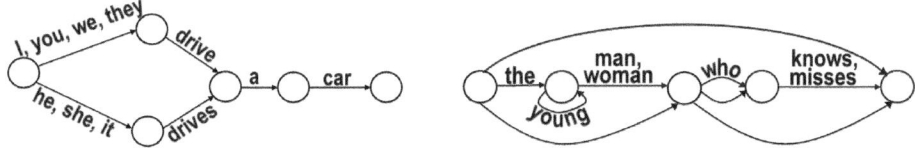

Fig. 1. Diagrams from the task of the Bebras competition used for generating English sentences and recognizing whether the given English sentence could be read according to the diagram

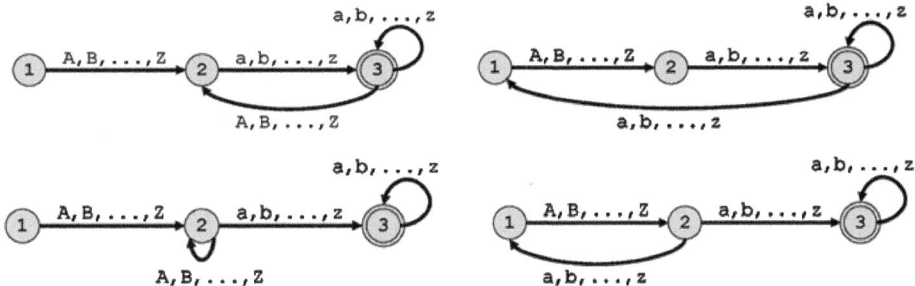

Fig. 2. Diagrams from the task of the Bebras competition used for recognizing login names

The tasks explained the meaning of the diagrams (how to interpret them), but still there was a lot of intuition needed (for example, one starts in the leftmost node and finishes in the rightmost node, when can one move from one node to another, etc.). Students were asked to explain where the starting node in the diagram is, and to make it clear that the starting and terminal nodes (later recognized as the accepting nodes) are important and have special meaning. The task in Fig. 1 was about generating or recognizing English sentences. The left diagram in Fig. 1 generates only a few sentences, but the right one generates infinitely many more of them. This was a striking discovery - that a *finitely described diagram can generate to infinitely many objects* (sentences). When teaching these concepts we made sure that the students understood the real power of these diagrams. The second task, Fig. 2, was to recognize the diagram that described the logins consisting of several parts, where each part starts with a capital letter followed by at least one lower case letter. Further examples included automata for generating necklaces and describing some processes. After everybody was comfortable working with such diagrams it was revealed that these diagrams represent an important TI computational model – finite automata. The diagram examples were used to explain what an alphabet is and what a word is, and we pointed out the differences and the similarities with the real life meaning of these concepts. We also noted that the next step in the diagram has always been uniquely determined. In CS we call this concept *determinism*.

The next step was to explore whether we could implement the diagrams in the programming language that the students learned in the programming modules. After some trials, students discovered that the diagram can be represented as a table, albeit two-dimensional. Each row in the table represents one node in the diagram. The columns represent arrows from this node (one for each arrow) and are labeled the same as the corresponding arrow. So the program simulating the diagram turns out to be one relatively simple cycle containing only one assignment: knowing the node we are in and the symbol at the arrow we have just read, we can find the next node in the corresponding row and column in the table.

Students also learned that it is very useful to have more than one representation of an automaton (and of any object in general) because different representations can be useful for different reasons. For example most people use a diagram representation of automata, but for computer programs a table representation is better.

We were now ready to formalize the concept of computation, machine, and the language it recognizes. The students were asked if they could specify what was done in the diagram examples in a more formal way. Could the computation be formalized with diagrams? With the students we specified the steps taken on the diagrams and thus defined the *computational step*. We then realized we needed a concept of *configuration* – the description of the momentary status of the computation. After some experimenting, we were able to say what computation is. The computation is accepting, when it is terminating in an accepting node in the diagram. At this point we could ask if for a given diagram it is possible to characterize all inputs it accepts – that is, *the language it recognizes*.

a   b

Fig. 3. a) Diagram of an automaton accepting words over the alphabet {0,1} with an even number of 0's b) Diagram of an automaton accepting words over the alphabet {0,1} such that the number 1 is divided by 3 and has a remainder equal to 1

The next task was to see if automata could be designed that would perform specific tasks. Along with the students we analyzed a very simple diagram with two nodes and a two-letter alphabet consisting of 0 and 1, which accepts words with an even number of 0's (Fig. 3a).

After successfully finishing this task we were ready to try to design a more real life automaton – a simplified coffee machine (nearly everybody in the class was familiar with the real coffee machine they used during each break). Our machine accepted only 50c and 1 Euro coins and dispensed tea, coffee and hot chocolate. The first step (not quite so easy) is to specify what the nodes are and what the alphabet of the desired automaton is. In this activity the students understood that designing the automaton is in fact very similar to programming. They realized quite immediately that according to their design decisions they needed different numbers of nodes, or to put it in another way, each design leads to a solution of a different complexity (Fig. 4).

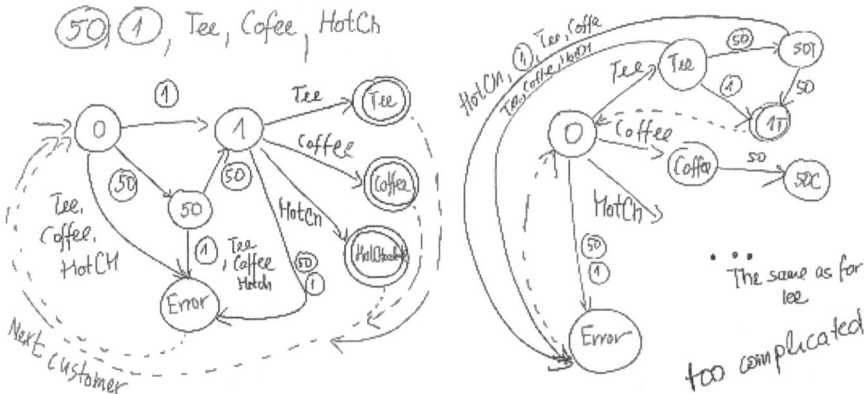

Fig. 4. Diagrams of the coffee machine

Another lesson learned was that it was far from clear that the designed automaton does what it was intended to do. We emphasized this fact and compared it with real situations we must cope with when we are programming. How is the program that is claimed to be the solution connected with the problem it was intended to solve? This is a very important but often neglected question.

The next level was the design of another "counting" diagram with three nodes and also with the 0 and 1 alphabet. This time it counted if the number 1 in the input divided by 3 has a remainder equal to 1 (Fig. 3b). This turned out to be quite easy.

We then asked if it would be possible to design an automaton which would simulate in parallel this automaton and the previous counting automaton (accepting the words with an even number of 0's). After some trials we found how to *simulate* two automata in parallel. This turned out to be more powerful than it looked at first sight, because it gave us a method for constructing correct automata, provided we can define the language which should be recognized as the union, the intersection and/or the difference of "simpler" languages. Again we emphasize the similarity with programming and the importance of a methodology that assures the ability to safely put together correct programs from smaller correct programs. Another important fact is the importance of the *decomposition* of the problem to smaller problems.

Towards the end of this module we returned once more to the diagrams in Fig 2. We then focused on the cases that were previously omitted as wrong, and in which the next step is not uniquely determined. Either there are two arrows with the same label or arrows were some labels are missing. With all our previous experience with the design and finding how things work it is not so surprising that the cases previously intuitively considered wrong can now be turned into correct ones. Of course, the understanding of some previously built concepts must be updated. However, this is the way scientists actually work The concept of computation must be updated to determine the acceptable computation. We introduce *non-determinism*.

In order to illustrate the usefulness of the nondeterministic finite automaton we used the automaton for a pattern-matching problem. This problem is familiar to most people who work with computers. For example, we want to find all files containing some text or find a given text on the web page or in a document. We can now explain that every nondeterministic finite automaton can be simulated by some deterministic finite automaton and therefore the deterministic and nondeterministic finite automata are equivalent (in their computational power). Unfortunately the number of nodes in the corresponding deterministic automaton can be exponentially greater than in the nondeterministic one. Fortunately this is not the case when we transform the nondeterministic automaton for the pattern-matching problem to a deterministic one which can work exactly as it works in real applications. The students appreciated this example of non-determinism because they are familiar with the problem, but did not realize it can be solved in this way.

Finally, the limits of automata were discussed with the students. We showed that the automaton has only very limited memory – in fact it has no variables, its whole memory is in the nodes (in the diagram representation). Therefore it will not be able to recognize the languages which require remembering some unbounded number (e.g. words with an equal number of 0's and 1's). Once it was recognized that the finite automata did not recognize every language, the natural question arose: Can we modify the computational model so that we can recognize languages that were not recognizable by finite automata? The answer is yes. It is possible to let students try to design such enhancements of finite automata. In the end we introduced the most general enhancement – the Turing machine. Our experience shows that designing a Turing machine is a highly elaborate and time-consuming process. We do not recommend doing this with in-service teachers in general, unless supported by excellent interactive software and a simulation tool.

When discussing Turing machines with the students it was emphasized that all concepts introduced by finite automata are still valid. Some definitions were modified, but all computations performed by finite automata can be accomplished. Moreover some computations that could not be accomplished before can be now. It is important to mention to the students that this will be the same with any other computation model they might eventually define and explore.

4 Observations

After completion of these modules, students were asked to fill in a questionnaire concerning the course. All activities from the lessons were listed:

1. playing with diagrams describing logins, English sentences, necklaces etc. using the diagrams as recognizers and generators of some words or objects.
2. formalizing the diagrams to finite automata.
3. discussing how to design an automaton that worked exactly as desired.
4. designing an automaton – a coffee machine for three kinds of drinks allowing the user to pay with two kinds of coins.
5. recognizing zip codes, car number plates, and national identification numbers.
6. formalizing the concept of a configuration, of a computation step and of computation as well as the accepting and rejecting computation.
7. discussing whether it is possible to recognize an arbitrary set with a finite automaton. (We found out that this is not the case and we explained intuitively why.)

8. extending the finite automaton model to a nondeterministic one.
9. As an example we showed how to find a substring in the text using a nondeterministic automaton.
10. showing the main idea of the equivalence of deterministic and nondeterministic automata, concerning the sets of words they recognize.
11. showing that an automaton can be described as a diagram, a program, a table or a regular expression.
12. extending the model of the finite automaton to the Turing machine, which can control the moves of the read/write head and rewrite the tape.
13. showing the difference in computation of a finite automaton and a Turing machine, concerning termination and accepting or rejecting.
14. discussing an extension of the Turing machine to a nondeterministic one.

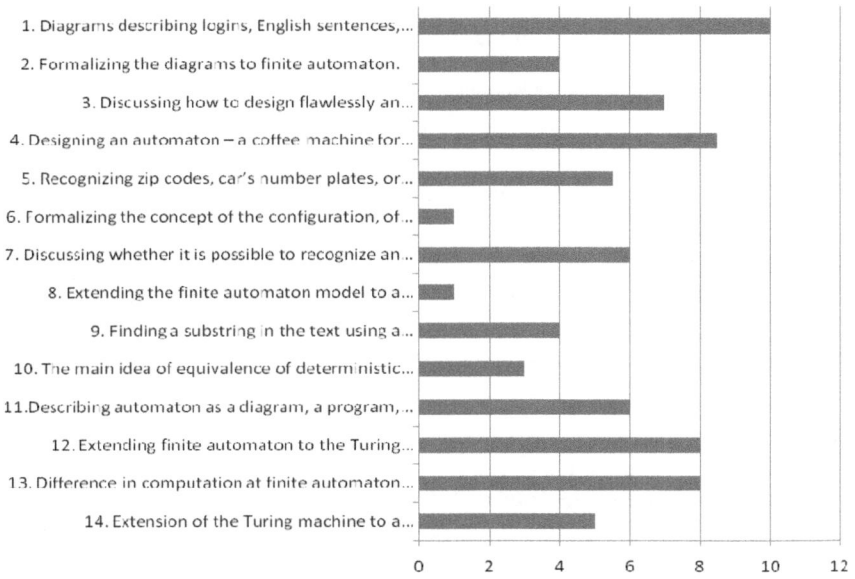

Fig. 5. Numbers of participants remembering listed activities

Students were asked the following:

a) List the number of activities you remembered,
b) List the number of activities you understood.
c) Have you studied some topics from TI in the past?

Out of 34 course participants, 13 answered our questionnaire. From those only one had studied TI previously.

Fig. 5 shows the responses to question a), (which activities participants remembered). Most answers had the form of a list of numbers identifying the remembered activities (as we had anticipated). One answer was "a few". Two answers were "all" or "I remember all activities, but mainly activities No.: …".

From the graph in Fig. 5 it is obvious that the most remembered topics were those connected with concrete examples (tasks from the Bebras competition or the design of the coffee machine). Less remembered were the more formal parts and generalizations (2, 6, 8, 10). A real surprise was that activities 12 and 13 (Turing machines and their computation) were mentioned relatively often, although we considered them as generalizations. Perhaps it is because they were covered towards the end of the course.

The second question concerning the understanding of topics learned was answered mostly with words and not with numbers (Fig.6).

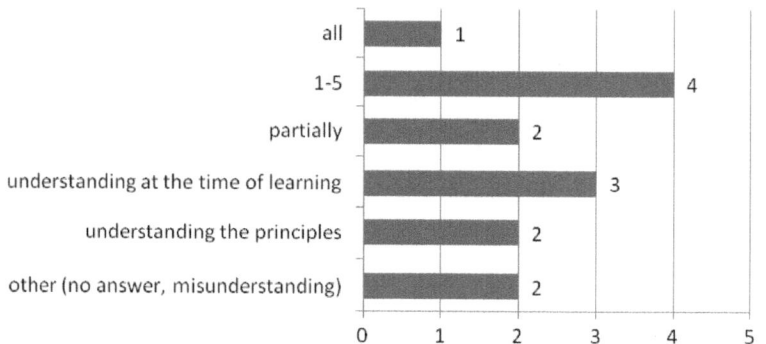

Fig. 6. Numbers of participants indicating their understanding

Answers were sorted into several categories:

a) **all** – only one answer,
b) **1-5** – four answers,
c) **partial** – two answers,
d) **understanding at the time of learning** – three answers
  o „*While listening to the lesson I had the feeling that I understood everything, but I did not remember it to the extent that I could reproduce it.* "
  o „*I understood everything during the lesson, but then I did not fix the new knowledge and did not devote any time to it.* "
e) **understanding the principles** – two participants - „*I understood the principles of all presented concepts, they were well explained, but I would not try to explore them in greater depth.* "
f) **other** (no answer, misunderstanding) – two participants. „*I cannot answer. I did not understand details, but it was great progress for me to hear about those concepts for the first time in my life. I am grateful for this, thank You.* "

Some participants wrote other messages:

• *"To be honest it was too hard stuff for me."*
• *"This topic was too remote for me as a lower secondary school teacher. I cannot imagine how to use it in lower secondary school. It must be very hard to mediate these concepts to us."*

Based on a short interview with two participants we concluded that some of the participants would appreciate even more examples, pictures and microworlds than we provided. The generalization of concepts was very hard and too abstract for some of the participants.

We conclude that in order to teach the introduction to TI to lower secondary teachers one must use many examples and progress very carefully.

5 Conclusion

There is not wide agreement on the best ways to teach future teachers of Informatics. There are two extreme approaches: A formal approach, where everything is taught very formally and rigorously using exact mathematical language, and a practical approach, where only immediately usable examples are emphasized. We believe that the right way lies (as usual) somewhere in-between those two extremes, especially if the mathematical background of the students is not strong. We are convinced that constructivist and constructionist approaches to teaching secondary school teachers TI will make their understanding of the important concepts of TI deeper and that as a result they will be able to use a similar approach with their students based on discovery and exploration in their teaching of informatics.

Acknowledgments. We would like to thank all the participants of the *DVUI* project for their inspiration during the two years that we taught at Comenius University in Bratislava, and thanks to Ivan Kalaš who initiated the idea of this paper, and to our families for their support.

References

1. Kalaš, I.: Discovering Informatics Fundamentals Through Interactive Interfaces for Learning. In: Mittermeir, R.T. (ed.) ISSEP 2006. LNCS, vol. 4226, pp. 13–24. Springer, Heidelberg (2006)
2. Hromkovič, J.: Algorithmic Adventures, From Kowledge to Magic. Springer, Heidelberg (2009)
3. Bebras, International Contest on Informatics and Computer Fluency, http://www.bebras.org/en/welcome
4. Forišek, M., Šišková, J.: Selected chapters on theoretical informatics 1 (algorithms) ĎVUI, ŠPÚ, Bratislava (2010) (in Slovak)
5. Andrejková, G., Krajči, S.: Selected chapters on theoretical informatics 2 (automata and Turing machines) ĎVUI, ŠPÚ, Bratislava (2011) (in Slovak)
6. Winczer, M., Galčík, F., Forišek, M.: Selected chapters on theoretical informatics 3 (Cryptography and randomized algorithms) ĎVUI, ŠPÚ, Bratislava (2011) (in Slovak)
7. Mittermeir, R.T., Bischof, E., Hodingg, K.: Showing Core-Concepts of Informatics to Kids and Their Teachers. In: Hromkovič, J., Královič, R., Vahrenhold, J. (eds.) ISSEP 2010. LNCS, vol. 5941, pp. 143–154. Springer, Heidelberg (2010)
8. Futschek, G.: Algorithmic Thinking: The Key for Understanding Computer Science. In: Mittermeir, R.T. (ed.) ISSEP 2006. LNCS, vol. 4226, pp. 159–168. Springer, Heidelberg (2006)

9. Gruber, P.: Bringing Abstract Concepts Alive. How to Base Learning Success on the Principles of Playing, Curiosity and In-Classroom Differentiation. In: Mittermeir, R.T., Sysło, M.M. (eds.) ISSEP 2008. LNCS, vol. 5090, pp. 134–141. Springer, Heidelberg (2008)
10. Weigend, M.: To Have or to Be? Possessing Data Versus Being in a State — Two Different Intuitive Concepts Used in Informatics. In: Mittermeir, R.T., Sysło, M.M. (eds.) ISSEP 2008. LNCS, vol. 5090, pp. 151–160. Springer, Heidelberg (2008)
11. Bell, T., Witten, I.H., Fellows, M.: Computer Science Unplugged, An enrichment and extension program for primary-aged children,
   `http://csunplugged.org/sites/default/files/activity_pdfs_ful`
   `l/CS_Unplugged-en-10.2006.pdf`

Informatics in Primary School
Principles and Experience

Andrej Blaho and Ľubomír Salanci

Department of Applied Informatics, Department of Informatics Education,
Faculty of Mathematics, Physics and Informatics,
Comenius University, 842 48 Bratislava, Slovak Republic
{blaho,salanci}@fmph.uniba.sk

Abstract. Informatics is gradually emerging, in different forms, at the first level in primary schools. We have seen that informatics is often included in other subjects but only at the level of using digital technologies. At other times, it is inspired by teaching methods in mathematics and computer science tasks that are addressed on paper only. We are concerned with the concept, form and teaching methods of informatics. In addition to scientific content, we analyze the teaching of informatics from the perspective of various educational theories. We have built our theory on the activities of students who work with specialized educational software. Our conclusion is founded on the experiences and consequences of introducing informatics education to teaching at the first level of primary school.

Keywords: primary education, informatics, curriculum, educational software, digital competencies, digital literacy.

1 Introduction

The attributes of informatics education at the first level of primary schools varies from country to country: work with computers is often scattered across different subjects with students remaining at the user level of digital technologies. In only a few countries does informatics constitute an independent subject. Slovakia joined this group of countries in 2009 when elementary informatics became an independent and compulsory subject at the first level of primary education. This arrangement followed ten years of experimentation with the teaching of computer science, during which time a variety of issues were addressed. For example:

- Why does informatics need to be a separate subject?
- Is it not sufficient for pupils to utilize digital technologies within other existing courses?

If we can manage to have informatics as a separate subject, then a series of other questions have to be dealt with:

- What should the content of informatics education be?
- What pedagogic methods are appropriate in the teaching of informatics education?

I. Kalaš and R.T. Mittermeir (Eds.): ISSEP 2011, LNCS 7013, pp. 129–142, 2011.
© Springer-Verlag Berlin Heidelberg 2011

The creation and proposal of a specific content and methodology of teaching informatics is a very difficult assignment. The complexity of this task is due to the fact that we are working with pupils aged 7-10 years, which means that we must understand a great number of varying factors. For example, we must take age into account, because we know the problem-solving capabilities of children at different ages, as well as what they are able to understand [1]. In addition, we must consider not only the global concept (curriculum, syllabus, textbooks for various classes, lesson scenarios, and the terminology used in tasks assigned to pupils) but also the pedagogical materials and appropriate educational software available to teachers. (The latter is not always obvious in our country).

2   Informatics Education Around Us – Analysis of the Status

It is now common for children to experience computers at home as well as in the early years of primary school. Even in pre-schools and kindergartens, computers are now frequently made available to children for play and learning activities [2].

2.1   Definitions

In order to better explain and compare the different approaches to the teaching of informatics, we will try to clarify some terms, such as 'digital technologies', 'digital literacy' or 'digital competence' in relation to informatics.

From our perspective, we have distinguished the following levels in the teaching of informatics:

- **The use of digital technologies = Use of DT** – this is the lowest level in computer studies. Here, the didactic priority is to teach a limited number of skills so pupils can use certain tools in order to enjoy working with computers.
- **Digital literacy = DL** – at this higher level of computer studies a pupil learns how to collect, evaluate, store, create, exchange, collaborate and communicate information. The priority here is to teach the correct use of applications.
- **Digital competence = DC** – this is an advanced level, which ensures that pupils can use information technology confidently and decisively for work, leisure and communication. Emphasis is placed on teaching positive attitudes towards computing, and on showing how computers can be used to resolve problems associated with the handling of information. [3]
- **Informatics** – exploring the basics of computer science. The priority here is algorithmic problem solving, the use of knowledge for the effective functioning of DT, problem solving, and an understanding of the range of influences on the current Information Society.

Note that each higher level covers the content of all the lower ones, but it exceeds the priorities.

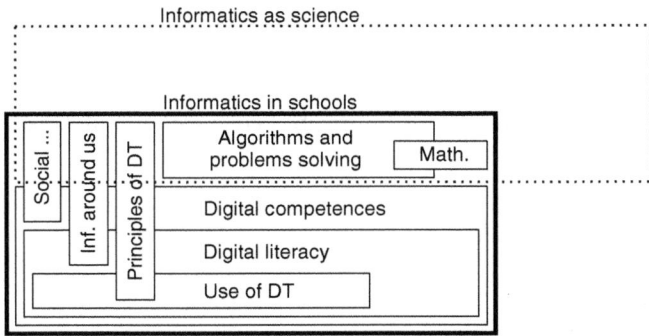

Fig. 1. The intuitive (and simplified) understanding of the relationship between DT, DL, DC and Informatics, discovered whilst analyzing the role of informatics in schools globally

2.2 Informatics in Other Countries

Interesting comparisons can be made with reference to the European study on Digital Competences [4]. Seven countries participated in this study: Slovakia, the Czech Republic, Portugal, Finland, Norway, Switzerland and Lithuania. Results show that all the participating countries, except Portugal, included some digital skills (digital literacy) in their educational system in primary schools. The study shows that digital competence is a core element in the national, primary curriculums of Slovakia, Norway, Switzerland and Finland. In addition to the Czech and Slovak Republics, where DT is taught as a separate subject, other countries are using cross-curricular ways of introducing computer skills.

There is a very interesting situation in Russia. The subject of academic computer science, particularly in secondary schools, has a longstanding tradition in this country. Over the last decade this subject has also penetrated into primary schools, and even into kindergartens. Russia, however, suffers from a huge discrepancy in the type of technologies made available to schools, with some primary institutions having no access to computers at all. The subject of computer science at primary school is not, therefore, focused on skills (in terms of the use of computers); priority is given to learning how to handle data structures, algorithmic problem-solving and logical thinking [5].

2.3 Informatics in Slovakia

An educational reform, launched in Slovakia in 2008, has brought great changes to the content of all subjects in primary and secondary schools. With this reform came the introduction of compulsory Informatics Education at the first level (grades 2-4) and computer science in the lower secondary level (grades 5-9). This was a completely new development with an innovative and beneficial vision of the educational syllabus. Before 2008 the national curriculum was designed by the

Ministry of Education, who decided what **teachers must teach** (an inputs-orientated curriculum). Now, conversely, the standards (ISCED) describe what **the student should know** (an outcomes-oriented curriculum).

Based on the 25-year tradition of Informatics education in Slovakia, informatics at primary and secondary schools consists of the following five collective thematic units[1]:

1. Information around us
2. Communication through DT
3. Procedures, problem solving, algorithmic thinking
4. Principles of the functioning of DT
5. The information society

These five themes are embedded at all levels of education, but at each level they have different goals and different demands. For example, while at level ISCED 1, the thematic unit for algorithmic thinking focuses on the introduction of basic concepts by means of controlling the movement of some character in an area; in level ISCED 3, pupils deal with the effectiveness of a particular algorithm and work with an actual higher programming language.

3   Forming the Conception of Informatics Education

We considered it necessary to find an answer to the following question: what are the aims of informatics education? We studied different approaches to the teaching of informatics in different countries and through analysis we discovered the various advantages and disadvantages of these approaches. Finally, we formulated for ourselves several requirements and principles, which we wish to pursue further.

3.1   Inspiring Approaches to Informatics Education

From the analysis of foreign educational programs and textbooks, we discovered two major approaches to the teaching of informatics. We called these approaches simply: 'user informatics' and 'mathematical informatics'. By clarifying the differences between these two approaches, we are able to reflect on their advantages and disadvantages. The following is an explanation of these approaches:

User informatics is focused on skills development, operating a computer and the use of software. Such education remains at the level of instruction and could be likened to ECDL where the pupils' skills are developed only to the point of knowing how to deal with and use DT. Such an approach has significant risks because instead of informatics, it pursues other objectives, such as: placing an emphasis on skills ("Type text as fast as possible ..."), using programs ("which command in the menu is used to ...") or teaching a sequence of steps leading to a certain result ("insert the image into

---

[1] It is interesting to note, that the ACM is beginning to consider a similar direction - see the new draft of the curriculum [6].

the text..."). Pupils are often evaluated for encyclopaedic knowledge ("List the rules of conduct on the Internet ...") or for compliance of aesthetic criteria while doing a task ("Nicely format the text ..."). In the last example, the teacher does not evaluate whether the student used the appropriate tools, but how nicely the student formatted the text – which is a goal of typography or graphic design). This means that instead of informatics criteria, criteria from other disciplines are applied. It is this user approach to informatics education, at the first level of primary school, which currently prevails in most countries.

Mathematical informatics is focused on solving computer science problems without using a computer. The main objective is not to teach pupils to work with a computer nor to gain digital literacy. Pupils solve problem-oriented tasks that are based on computer science: getting to know the idea of sequences ("Arrange beads on a necklace so that ..."), getting familiar with the combinatorial principles ("How many objects have property ...") , learning to design and read algorithms ("Write a program for the robot so that ..."), reflecting on strategies (" What might be the outcome of the game ...") or getting to know structures ("Which texts are hidden in the tree ..."). Tasks are designed so that they can be solved on paper, without a computer. However, if computers and specialized applications (so-called micro-worlds) are available, pupils have the advantage of programs that can help them solve problems, offer variations, or evaluate a solution. A mathematical approach to teaching computer science has, in our opinion, several disadvantages. Since, with this approach, working with a computer is postponed until later grades, there is a risk that pupils acquire digital literacy unnecessarily late, and they do not learn some things at all. In addition, pupils, parents and even teachers may not (possibly) understand why the children only deal with mathematically oriented tasks. This approach places much more stringent demands on both the teacher and pupil.

3.2 Our Vision of Elementary Informatics

Our Elementary Informatics will be, to some extent, a combination of the two previously mentioned approaches with an emphasis on choosing the very best ideas from both. We want our pupils to learn how to work with computers right from the very beginning while, at the same time, they are solving informatics/computer science problems.

   We know that it is not easy for young pupils, who have no prior experience working with computers, to coordinate the movement of the hand whilst they follow the mouse cursor on the screen. Even more difficult is the action of placing a mouse cursor on the correct area of the screen, and then to click on this area <u>without</u> moving the mouse at the same time. This is a basic skill of **using digital technologies** that all pupils need to learn. We want to teach this development of fine motor skills by engaging pupils with playful computer activities - but at the same time we want these activities to contain various **informatics elements** (Fig. 2).

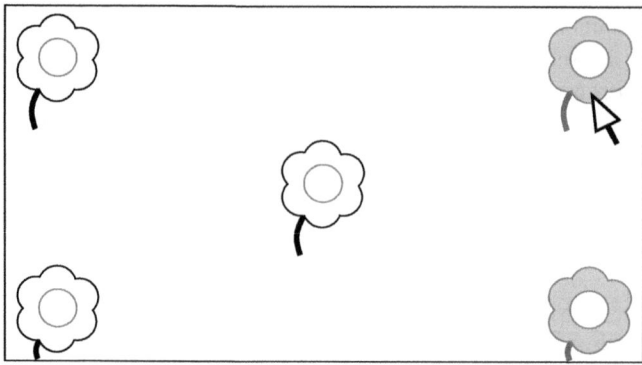

Fig. 2. In this activity "it is necessary to click on all the flowers in the corners first, and to leave clicking on the central flower until the end". This activity is for pupils of age 7. Whilst on first sight it appears that only computer skills are being taught, the task also contains simple rules, which the pupils must follow – these are sequencing and abiding rules that are primary computer science concepts.

In activities that are aimed at the development of digital skills, we will incorporate computer science content, even if pupils are being taught to work with a keyboard, graphics, texts, tables, sounds, the internet etc. We are also trying to implement a similar philosophy conversely. This means that pure computer science topics can be realized in a way that develops a pupil's digital competency. For example, if we teach topics like logic, combinatorics, algorithmic problem solving, or informatics structures, we will try to present them in a playful form using computers (Fig. 3).

Fig. 3. Kitty wants to build a tower as shown, but she does not know in what order to pick the blocks from the box. Help her by lining up the blocks in the correct order." This activity is for 7-year-old pupils. A computer scientist will very quickly recognize, through this activity, an introduction to algorithms (a simple design formula). Pupils are, moreover, taught how to drag objects using the mouse (a skill which is at the use of DT level).

In our view, informatics should contain the following:

Development of Digital Skills + Development of Informatics Thinking

3.3 The Emphasis on Pupil Activity and Amusing Tasks

From the previous examples one can already see that we are placing great emphasis on activities that are fun to do. To some this approach may seem the obvious one to take, but to others it may appear frivolous and too playful. We therefore must address both points of view.

During the formation of our theories we noticed that such a playful approach is not at all obvious to many people. Many authors of textbooks and children's books about computers (as well as teachers) do try to pass on information to their pupils – at first they focus on explaining the theory and only then go on to give examples and demonstrations. In our investigation we have seen this approach at work:

- At the beginning of the hour-long lesson a teacher will explain to the pupils what a computer is, its different parts and how it works. However, for many pupils, this might be the first time that they have ever encountered a computer.
- At the start of the topic, Working with Graphics, a teacher describes the graphical environment, the buttons, menus, and the commands in the menus. The student can only begin to draw when the description is finished.
- A class about algorithms begins with an explanation of the properties of algorithms, the stages of problem solving, and how large companies develop software from this perspective. Only after the teacher's explanation can a pupil begin to write their first program and learn more about programming.

Such a traditional, authoritarian approach has a long tradition in Slovakia. However, we consider it inappropriate, since pupils are treated as passive recipients of information. They sit and listen to the teacher or watch a presentation, and this kind of lesson usually ends up being mostly boring for them. It is risky for a teacher to talk about a topic with which his pupils have little or no experience. For example, when a teacher is talking about the characteristics of algorithms before the pupils have programmed for the first time, that teacher is skipping five levels of Bloom's taxonomy: the consideration of the properties of algorithms (which is the highest level) are then handled by pupils who do not have any programming experience nor even a knowledge of algorithms, which is considered the lowest level of Revised Bloom's taxonomy. Consequently, a pupil's knowledge is very shallow and has a short lifespan. As the pupils begin to formulate their own solutions to a task, they often realize that they did not understand the teacher's presentations.

We want our pupils to perceive work performed during an Informatics lesson as fun and as a game – but having fun is not the only goal of this approach. From the perspective of Informatics, pupils get to know **serious Informatics concepts** such as: propositional calculus, combinatorics, structures (sequences, tables, trees, and graphs), algorithms (writing, reading, debugging, and efficiency), principles (coding, functioning of devices), etc. We have to distinguish between playing games on the computer, into which informatics teachers like to slip, and actual problem-solving activities.

3.4 We Rely on Theories

In our hypothesis, we rely mainly on **constructionism** as a theory of learning. Put simply, this theory says that knowledge cannot be passed on through a passive reception of ideas and facts. Knowledge must be constructed by the pupil. They must discover it for themselves with the help of a facilitator in the form of a person, a book, a film etc. – and, most productively, through an activity that *they* perform [7]. We have had many good experiences with constructionism in teaching informatics and, therefore, want our approach to teaching to be based on **pupil activities**. In this way pupils learn to solve a number of tasks and problems independently. We want to suppress the instructive approach to teaching as much as possible.

Of course, it will have a major impact on the role that the teacher has in the classroom. The subsequent change will not be easy. Teachers will have to transform their role from that of a person who stands before their pupils whilst explaining topics to them and then testing their knowledge, to that of someone who helps, advises, and guides their pupils appropriately. If a teacher does need to explain something to the pupils, then we strongly advocate that they explain only the **necessary minimum**.

However, we want to avoid so-called naive constructionism. This approach leaves the pupils without proper leadership, to haphazardly grope their way through problem solving, because the tasks they have been set are poorly conceived. Our activities, therefore, have to be precisely and very shrewdly designed. We have several instruments which help us to avoid bad practices.

The theory of **constructivism** [1] provides us with levels of cognitive development that informs us about what pupils at a certain age are able to understand and how they formulate their understanding. Our pupils are at the beginning stage of **concrete operations**, so they can think logically and solve problems which concern only specific objects. They do not strive to solve tasks that are based on formal definitions.

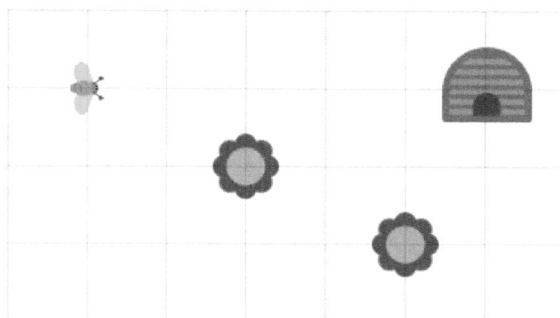

Fig. 4. Appropriate motivation is extremely important for younger pupils. Therefore, our tasks are embedded in stories, for example: "Draw a path for the small bee to visit all the flowers and then to finish in the beehive." From the perspective of Informatics this is a simple graph problem.

Furthermore, we believe that informatics knowledge is formed gradually in a number of stages. We, therefore, follow every single stage of the cognitive process [8]:

Motivation → Collecting of experiences → Abstraction →
Knowledge → Training of knowledge

It appears that pupils understand new concepts only after they have had numerous experiences with them. They then draw comparisons among the experiences and put everything into a context. Teachers can very effectively help to stimulate motivation in children and provide situations for them to collect their own experience.

In order to make the tasks and texts for pupils understandable, we find it necessary to prepare our own dictionary of the words that pupils should know or progressively get to know. We also try to imagine (and design a model of) what pupils already know.

3.5 Conclusive Fine-Tuning

It was absolutely necessary for us to subject informatics topics to rigorous **analysis** in order to find a place for them within the elementary curriculum for teaching computer science. We, therefore, examined textbooks for other subjects, as well as foreign textbooks on Informatics.

In addition, we **consulted** with a selection of practicing teachers about our ideas. Such cooperation proved to be extremely rewarding. The teachers helped us to adjust the level of difficulty in our tasks so that they are more appropriate for pupils. Moreover, they helped us understand the many subtle relationships between teachers and pupils, relationships regarding subject areas, regarding the sequencing of topics, and the ages of the children and their motor skills and ability to understand things.

We discovered a **number of various issues** to be aware of in order to prepare suitable tasks that reflect the learning stages of the pupils. For example, if we want to teach 2nd grade pupils to use mouse clicks competently, we should not set them a task that specifies "click on the letters in alphabetical order", because alphabetical ordering is not taught until the 3rd grade. Similar issues exist not only among subject areas but also in elementary informatics itself. For example, we cannot begin to specify a task such as, "Write the following URL address in the browser window...", if a large group of pupils does not know how to type.

After long consideration, we decided on the division of elementary informatics into the following thematic units:

- – Working with text,
- – Working with graphics,
- – Working with multimedia,
- – Working with a story,
- – Structures in informatics,
- – Internet and communication,
- – Problem solving and algorithms,
- – About a computer,
- – Information Society.

We analyzed each thematic unit and we **clarified**:

- – what is the educational objective,
- – what activities we need to implement with the pupils,
- – what informatics competencies we need to develop through each activity,
- – what tools we will use in teaching,
- – which terms pupils need to learn,
- – in which grade do we need to implement activities.

We have designed a uniform structure, which we have written down using tables similar to those in Table 1 (below). A comparable series of tables was created for each thematic unit in which we methodically sorted activities according to their different dimensions.

Table 1. A part of the table that relates to the inclusion of IT structures in the teaching of elementary informatics. We also created similar tables for other topics:

Activities	Informatics capability	Tools	Terms	Year
Sequence and completion (geometric shapes, letters ...) generating according to certain rules finding errors	understanding of the sequence properties object manipulation interpretation of rules of a given sequence	graphical editor, or a specialized children's program	sequence missing, unused element true, false	2-4
classifying of things and objects according to certain characteristics	the recognition of common properties and assigning things to categories	graphical editor, or a specialized children's program	group, belongs to, not belong to	2-4
making of a frequency table and a bar chart reading of a specified diagram (chart) gathering information from classmates (about animals) and creation of a frequency table	become familiar with tables, making and reading tables and bar chart	graphical editor, or a specialized children's program	table bar chart maximum, minimum	3,4
etc.	...	...	...	...

4   Outcomes, Observations and Evaluation

The first consequence of our studies is the development of an **informatics education textbook** for the 2nd grade [9]. There are approximately 200 assignments, or activities, in this textbook - all of which need specific software for a given activity.

The new **national project** called "In-service Training for Informatics Teachers"[2] (ITIT) which began in Slovakia in 2009 aims to develop and implement pilot training for informatics teachers. Several hundred teachers (which is only about 10% of all such teachers in the country) have consequently received basic training in how to teach/learn subject matter pertaining to informatics. These teachers developed digital capabilities, teaching competencies using digital technologies, and skills in instructing pupils about elementary informatics at primary schools. These are all very different and challenging skills. As project designers, we had the task of developing educational materials for teachers on digital literacy and the didactics of informatics education: the most interesting of undertakings.

Because of this project we met with many teachers from primary schools. We learnt about their attitudes and received their feedback and observations.

---

[2] In Slovak: Ďalšie vzdelávanie učiteľov informatiky (DVUi), dvui.ccv.upjs.sk

4.1 Teachers' Attitudes

We learnt that the majority of teachers have initial difficulties with the application of technologies - which results in false expectations about the kind of problems that students may have when first introduced to digital technologies. Many teachers **do not understand the level of difficulty** in passing on skills and knowledge and, therefore, teach too much content (often didactically inappropriate). They frequently use unsuitable tools, insist on unimportant details, and fail to understand basic concepts and principles about the function of digital technologies. The following quote by a teacher illustrates this situation: „Until we finished the ITIT training we had a completely different idea about how to teach informatics."

Since teacher development in DT started in Slovakia such a short time ago (2009) , only a minority of teachers understands DT, are aware of the risks of using DT (e.g., viruses) and understand the principles of personal data protection. We observed, also, that teachers frequently avoided the topic of „Algorithms and problem solving". However, their opinions may change when they begin to observe which activities are most enjoyable to their pupils.

The ITIT project has had a great impact on teachers. Many have been left with a positive attitude towards the training that they received. Moreover, many teachers who have completed ITIT training now look a bit more critically at colleagues who have not been trained.

It is rare that primary school teachers have a proper informatics background. When they do have it, there is a good chance that they also have high standards of teaching. There is, however, the irony that such teachers may overload their pupils with too much inappropriate content.

4.2 Software and the Internet

Our approach assumes that students will work with computers and appropriate programs. Each activity should be associated with a specialized program designed to help the pupil solve the assigned task.

At present (at least in Slovakia) there is a shortage of **programs for children**. Examples of desirable software include: a suitable text editor, a spreadsheet editor for children, a children's tool for working with sounds, etc. The reasoning behind this is that environments for children need to look completely different from those designed for adults (for example, we can compare the children's graphic editor, Revelation Natural Art, with an environment suitable for professionals, e.g. Photoshop or GIMP).

A separate group of problems deals with teachers' attitudes to software. Teachers often take existing software as an authoritative product 'given from above' and because it is installed on their computer, it must be used by pupils. In addition, many teachers want students to use the broadest possible range of instruments, often of dubious quality, arguing that "the students get bored when working with one program" or that "students need to amass as much experience as possible of different programs". These are the main explanations for why pupils, for example, end up using five different graphic editors. We consider this highly inappropriate for children in this age group.

The teacher might not be aware that a pupil working with unsuitable software can get de-motivated, and even formulate misconceptions about the functionality of DT. We often encounter a recurring attitude amongst teachers, parents, those without experience in computer science, as well as the professional community, that open-source programs, freeware, and shareware need to be utilized as much as possible. The reasoning for this is that free software reduces "the financial budget of the school" and pupils are also, at no extra cost, able to install the programs at home. But this software is often didactically very inappropriate. The authors of such programs are most often professional computer scientists who are very good programmers, but who have no pedagogical experience in education. They often do not take into account the real users - that is, the technically inexperienced teachers and the 6–10 year-old pupils.

A big problem is the use of office suite packages for teaching pupils. In terms of criteria for educational software, these office suites are wholly unsuitable for the classroom. We believe that office programs violate most of the criteria by which we assess the suitability of a teaching aid, for example:

– the number, size and layout of controls,
– the ease of the dialogue - working with files, adjusting fonts, paragraphs ...
– the number of possibilities for different options – a variety of font types, numerous styles, a selection of colours,
– the operations - inserting images, tables, fancy fonts, placing them (wrapping).

If we compared this software scenario with a similar case in any other subject taught at this school age, we would not find teachers using products with their pupils that were intended for adults. It seems that only **in informatics do we find poor understanding of the risks associated with the use of inappropriate tools**.

There is no doubt that the majority of internet content is intended for adults with much of it being inappropriate, or even detrimental, for children. The educational institutions do almost nothing **to protect children against harmful content**. If there is any consideration about this problem, then it is voiced only through a general statement about the need to protect children from accessing inappropriate websites. Unfortunately, there is practically no initiative for addressing this problem by creating a **web for children**, which would include enough resources to make accessing adult sites unnecessary.

4.3 Attitudes of the Public

The majority of the public, even though they do not have any computer science education, is eager to give opinions on the integration of DT - particularly in education. Teachers may encounter this fact when they communicate with parents who have learnt about DT, and how it is supposed to be taught, from different media. It is interesting to note that no other school subject (e.g. mathematics, science, language studies) has caused so much interest amongst the general public and spawned so many different opinions. It is often obvious on many levels that an author/programmer did not receive any computer science tutoring during their own schooling and, therefore, has no grasp of the contemporary educational system. We often encounter inappropriate advice about how software, computers and the Internet should be used in schools or how a teacher should teach, classify, evaluate, and so on. This makes it far more difficult for teachers now than in the past (for example, before the internet was invented), to explain to parents the objectives behind introducing DT into the classroom.

The professional public (e.g. experts in computer science - practical or academic) is a very interesting group to take into consideration. They like to speak about informatics content in elementary and secondary schools. Informatics specialists also often recommend that students should use the most popular technology at the time: modern software packages, the latest technological fads, the most successful professional approaches and so on. Such experts seldom realise that **schools educate pupils for the future** – our students, after all, will begin their professional lives sometime in the next fifteen years or so. Therefore, it is more important to teach about general principles than to promote current technological innovations. Some academic computer scientists do have the courage to speak about the content and methods that should be used when teaching informatics in schools. They recommend that students should focus on mathematical foundations for algorithmic problem solving. They recommend this instead of students learning how to control applications for processing various types of information, working with the Internet, using DT for communication, learning the basics of programming concepts using computers in a playful manner, learning about the social aspects of computer science and so on. Actually, they do not consider this as informatics. We commented in the previous chapter (Mathematical informatics) why such an attitude is inappropriate.

5 Summaries

A lot of issues must be considered when designing a new curriculum: the topics to be included in informatics, the relationship to other subjects and the didactics of informatics in the first stage of the primary school. An important outcome of our studies is finding that adequate software tools must be developed for students. Our experience shows that only a limited number of people at present can understand the concept of good educational software and are able to evaluate or even develop such tools.

We discovered that most primary teachers do not understand what informatics means, what the objectives are for teaching informatics or what themes it addresses. Almost all teachers think that the curriculum is comprised of how to use digital technologies as a tool. We have shown that consistent design, analysis and the writing of textbooks is not enough to make informatics education successful in schools. It is very important that teachers receive appropriate methodological materials and, even more essential, that they receive adequate training.

References

1. Rybár, J.: Úvod do epistemológie Jeana Piageta. In: Introduction into Epistemology of Jean Piaget. IRIS, Bratislava (1997) ISBN 80- 88778-43-3
2. Moravčík, M., Pekárová, J., Kalaš, I.: Digital Technologies at Preschool: Class Scenarios. In: Proc. of 9th WCCE: IFIP World Conference on Computers in Education (2009) ISBN 978-3-901882-35-7

3. Review of National Curricula and Assessing Digital Competence for Students and Teachers: Findings from 7 Countries, Digital Skills Working Group, European Schoolnet (2010)
4. Recommendations of the European Parliament and The European Council about the key competences for the life-long learning. Official Journal of the European Union (December 30, 2006)
5. Semenov, A.L., Rudchenko, T.A.: Series of textbooks of Informatics, Institute of new technologies, Moscow 2004 (2005) ISBN 5-09-012563-5, ISBN 5-09-012562-7, ISBN 5-09-013872-9, ISBN 5-09-013873-7, ISBN 5-09-013876-1, ISBN 5-09-013877-X
6. New CS Curriculum Standards (Draft for Public Comment), http://csta.acm.org/includes/Other/CS_Standards.html
7. Ackermann, E.: Constructivism(s): Shared roots, crossed paths, multiple legacies. In: Proc. of Constructionism 2010. Comenius University, in association with The American University of Paris, Bratislava (2010) ISBN 978-80-89186-66-2
8. Hejný, M., Kuřina, F.: Dítě, škola a matematika. In: Children, School and Mathematics. Praha, Portál (2009) ISBN 978-80-7367-397-0
9. Blaho., A., Salanci, Ľ.: Informatická výchova pre 2. ročník ZŠ, Bratislava (2010) ISBN 978-80-89375-17-2

Teaching Programming at Primary Schools: Visions, Experiences, and Long-Term Research Prospects*

Giovanni Serafini

Department of Computer Science, ETH Zurich, Switzerland
giovanni.serafini@inf.ethz.ch

Abstract. The key contribution of computer science to general and school education relies on the concept of Computational Thinking. Teaching programming in Logo at the primary school is an appropriate didactic approach towards Computational Thinking, it permits to embed Computational Thinking into a spiral curriculum at a very early stage and should enable specific transfer to related school subjects. The paper describes our concrete experiences in teaching programming in Logo at Swiss primary schools, reflects on didactic visions and consider prospects for long-term empirical research.

1   Introduction

Computer science is nowadays omnipresent in real life. University programs in computer science were already introduced in the late sixties of the last century and are actually well established. It is to be considered a kind of paradox that despite of its everyday relevance and its importance for university education and research, efforts aiming at introducing a dedicated computer science school subject evoke controversial discussions among school communities as well as among the scientific community [10]. A missing general consent about goals, teaching topics, and even about the scope of computer science and the delimitations to related disciplines characterizes the debate.

Jeannette Wing brings order into the discussion declaring Computational Thinking as the key contribution of computer science to general and thus to school education [11]. According to Wing's vision, we believe that Computational Thinking is an attitude as well as an extensive framework of concepts, abilities, and skills young people should learn in school. We are convinced that learning programming on an adequate level of abstraction is a very effective didactic approach to Computational Thinking, independent of the age of the pupils.

In this paper, we focus on our school projects aiming to introduce primary school pupils to Computational Thinking by teaching them how to program. The pupils usually attend grade 3 to 7 and are roughly between 8 and 13 years

---

* This work was partially supported by the Hasler Foundation.

I. Kalaš and R.T. Mittermeir (Eds.): ISSEP 2011, LNCS 7013, pp. 143–154, 2011.

old. During the last 6 years we were allowed to teach several hundred children programming in Logo, during on-site school projects as well as in dedicated events directly at our university.

The paper is organized as follows: Section 2 addresses the nature as well as the scientific fundamentals of Computational Thinking. Section 3 focusses on the adopted didactic approach while teaching programming at the primary school: we discuss the didactic and technical requirements for a programming language for very young pupils, present the vision and the didactic approach we follow, and further give a concise overview of the teaching materials we developed. In the following section, the structure and the timeline of a typical school project are shortly addressed. In Section 5, we reflect on general experiences and learning achievements during the school projects. Section 6 highlights research prospects dealing with Computational Thinking and with starting programming courses at primary schools, and finally outline prospects for empirical research.

2   Addressing the Nature of Computational Thinking

2.1   Computational Thinking Is Unique to Computer Science

In a seminal contribution, Jeannette Wing highlights Computational Thinking as a new thinking paradigm, which enables innovative ways for addressing problems in everyday life as well as for approaching research subjects in apparently completely unrelated scientific fields. Wing assesses that everyone, not only computer scientists, would be eager to learn and to use it [11].

Juraj Hromkovic explains in detail, that computer science has aspects of mathematics, natural sciences, technics and even of philosophy, but is not exactly included in one of these. He highlights that computer science is an independent scientific discipline which formally studies ways to automatically (algorithmically) solve problems, and assesses that the concept of algorithm should be considered as the first axiom of computer science. Since algorithms are the core of computer science, Computational Thinking is the key contribution of computer science to general education [3,5].

According to these two very similar definitions, algorithms are unique to computer science, and computational (or algorithmic) thinking can only be learned in a dedicated computer science school subject.

2.2   Children Should Learn Computational Thinking for Life

Computational thinking is a scalable concept whose extent can be adapted, depending on the abstraction skills and on the prior knowledge of the involved actors. We think that Computational Thinking can be introduced and taught on an appropriate and adequate level of abstraction at all school stages, and that it is well suited to be taught in a cognitive actively form of learning, within a spiral curriculum. We are firmly convinced that learning how to program represents a scientific sound, but very effective didactic approach to Computational Thinking, independent of the age of the pupils.

We believe that Computational Thinking is nowadays essential and that children should learn it for life. We therefore promote the idea that a computer science subject relying on Computational Thinking should be mandatory at every school stage, including the primary school. We do not pursue the goal to make a computer scientist or a software engineer out of each child, but we assume that introducing programming at a very early school stage might help increasing the interest in computer science and MINT subjects[1]. We hope that therefore, on a long-term perspective, more school-graduates would consider to enrol for a university program or to start a career in a MINT field.

3   Didactic Concept and Teaching Approach

3.1   The Quest for the Programming Language

We believe that Logo [9] still is one of the most adequate programming language for beginners, particularly for classes at primary school. Logo is a mini-language and, in contrast to general-purpose programming languages, it was explicitly developed for teaching programming [4]. In order to write and run Logo programs, beginners do not need to learn a consistent language subset and do not have to deal in any way with the intrinsic complexity and the particular features of a larger, general-purpose programming language [1]. Logo permits the user to focus on a very reduced set of instructions with an adequate, clear syntax. In contrast to sophisticated programming environments for other, well established mini-languages, simple Logo editors do not rely on a click-and-drag approach and allow the pupils to type the instructions by themselves. This permits them to care for correct syntax and dramatically reduces the cognitive overload caused by an unmanageable list of instructions or by an overcharged graphical user interface.

Moreover, the open-source programming environment XLogo [7] satisfies all our most relevant expectations. It is free of charge, it runs on multiple platforms, it does not need an internet connection at run-time and runs quite well on slow computers. Schools are therefore able to simply deploy it, even in older computer rooms, teachers as well as pupils and their families are allowed to use it without further constraints on their private computers.

3.2   Didactic Approach and Teaching Materials

The German textbook *Einführung in die Programmierung mit Logo* [4,2] is a detailed introduction to programming in Logo permitting to intensively teach classes of different school stages on a regular basis. The contents of the first part of the textbook are well suited for teaching programming at primary schools.

School projects aiming to introduce children to programming usually have very restricting time limitations. We therefore decided to develop compact, ad-hoc-teaching materials for primary schools adhering to these constraints which

---

[1] MINT is an acronym for mathematics, informatics, natural sciences, and technics.

still rely on the didactic guidelines from the mentioned textbook. The teaching materials are meant to be the ideal support for our school projects and have further made been available online, free of charge, for interested schools and institutions [6].

The teaching materials are organized in six different lessons and do not require prior knowledge in programming.

3.3   Lessons 1 to 4

The first lesson introduces the concept of a **program as a sequence of simple computer instructions**. The pupils learn the exact meaning of the four basic instructions *fd*, *bk*, *rt*, and *lt* which permit to move and to rotate the turtle, as well as the instruction *cs* used to clear the screen. They learn that the computer does only understand clear, unambiguously formulated instructions, and they further get introduced to the simple, but very effective XLogo-programming environment.

Lesson 2 focusses on the concept of a basic loop with a constant, a priori known number of iterations. The children are lead to discover the need for a way to automatically repeat a specific code segment, and learn the exact, but not yet very intuitive syntax of the *repeat* instruction. They start therefore writing shorter, clever programs which reuse code they already wrote and directly influence the sequential execution flow. At this stage, children do not need to be introduced to the concept of a variable.

Lesson 3 introduces the approach of modular design of programs, which is not unique to computer science and represents an ideal bridge to other technical disciplines. First of all, the pupils learn that programs mostly need a name and that a named program can be nested in a larger program. In a small project, pupils have to draw a town consisting of identical streets of exactly the same house type. The children recognize that modularity is a very powerful concept, permitting to conceive and produce programs solving complex assignments. They understand that modular design simply reuses existing programs in a very elegant way. The assignments need the two additional instructions *penup* and *pendown*.

Lesson 1 to 3 allow us to highlight some characteristics of the chosen didactic approach and of the teaching materials:

- There is no systematic direct instruction of the children by a teacher. The pupils are expected to work autonomously, relying on the teaching materials and on the interactions with the tutors.
- The theory blocks and the texts are very short, the language is simple, but the terminology is precise.
- Pupils are able to write and execute their first program within minutes. They only need to know a few instructions.
- The exercises of the first two lessons are simple and permit the pupils to familiarize themselves with the programming language and the basic concepts we introduced.
- The exercises focus on small algorithmic problems: the assignments are concise and the expected solutions are compact. Pupils are even able to verify

their solutions by themselves, on the screen, without a formal feedback of a teacher. If not, they can autonomously start debugging the code.
– A didactic sound approach allows to rapidly introduce basic concepts such as program, loop and modular design.

The fourth lesson deals with drawing regular polygons and circles. The new instruction *setpencolor* permits to change the color of the pencil. In this lesson, no new programming concepts are introduced. The pupils are mainly expected to gain deeper routine in programming by solving assignments for which they generally have prior knowledge from geometry and mathematics.

3.4 Lessons 5, 6 and the Concept of a Constant Parameter

Lesson 5 introduces a substantially simplified concept of a variable in form of a constant parameter. Relying on a placeholder abstraction and lead by the teaching materials, pupils learn that programs are usually able to solve a class of problems whose instances only differ in the value of one or more specific parameters. The initial example focusses on drawing a square. The exercises permit to vary the figure the pupils have to draw and to change the number of parameters. This approach allows to efficiently mask the hardware layer and therefore to reduce the intrinsic complexity of the concept of a variable to an appropriate level for primary school kids.

In Lesson 6, the pupils learn to combine parameters with modular design. They write challenging programs expected to pass parameter values to subprograms, integrating all the basic concepts and the instructions of the lessons 1 to 5.

The didactic approach behind lessons 5 and 6 is similar to the approach we chose for the first four lessons. Even if contents and exercises are now more complex, the pupils are still writing small programs based on very few instructions, in a simple and effective programming environment. The pupils can still learn, program, test, and debug in an autonomous way, while tutors assist them.

4 Programming Projects at Primary Schools

4.1 School Projects

Practical organisational constraints impose a very similar but flexible plan for each new school project. A typical school project currently consists of 12 to 16 units of 45 minutes each, which are usually taught in 3 to 4 half-day blocks. Classes are split into parallel groups of up to 12 pupils. Two tutors and usually one class teacher supervise each group. Each child uses an off-the-shelf-netbook, learns and programs autonomously and is therefore implicitly able to adapt his pace. Each school project ends with a programming contest which give us a more concrete impression of the learning achievements of the pupils. A closing act give us a feeling of the attitude of kids, teachers, school board and political authorities with respect to programming classes.

4.2   Class Teachers and University Tutors

We explicitly involve the teachers at the host-schools and expect them to actively support their classes during the school project, but we have no particular requirements with respect to their prior knowledge in programming. According to the mission of the primary school, classes are usually taught on a one-teacher basis and teachers are therefore required to integrate wide-ranging school subject competence with deep pedagogical and didactic skills. Their education in computer science mainly focusses on mastering computer applications which support the preparation and the documentation of their lessons. Consequently, the first project stage consists of training the teachers. This usually happens on-site, during a half-day workshop, relying on exactly the same didactic approach and the same teaching materials we later use with their pupils.

The school projects still involve tutors from our university. The composition of the team varies and comprises faculty members, lecturers, postdocs and PhD candidates, as well as several undergraduate computer science students. Female and male tutors are usually equally represented.

5   General Experiences

In this section, we present selected major observations from our school projects. These observations are common to most of the school projects and enable us to summarize our experiences in a general, meaningful way. Furthermore, we consider the results of a survey among pupils and teachers taking part in recent projects, present the assignments of a programming contest, and comment the results. The reflections generally focus on didactic issues to be additionally addressed and eventually empirically investigated.

5.1   Major Observations

- During each project, the pupils need up to 30 minutes at the beginning of day 1 in order to get used to learning autonomously.
- Pupils get rapidly used to the Logo syntax. Typing the code by themselves is not a particular issue.
- The pupils are really dedicated. They visibly like programming and are still proud to show their progress and the solutions to the assignments. They usually have to be forced to leave the classroom during the mandatory breaks.
- The kids always try to improve their solutions. The are not satisfied with a weak solution.
- Girls are as engaged and achieve the same goals as boys.
- Starting with lesson 4, the cognitive complexity of the taught contents increases. Pace differences among the pupils increase, too. Since the pupils are working autonomously, tutors do not have to assist the whole group but are able to actively support kids having particular problems or looking for being challenged.

- Teachers are really motivated, even if they do not have any kind of prior knowledge. They enjoy to learn programming, like to have well-prepared teaching materials they can later use and like the exchange of insights about programming and didactic issues with university staff and tutors.
- The teachers are not necessary faster in learning than the kids.

5.2   School Project in Attinghausen

A recent project took place a the primary school of Attinghausen, in the Swiss Canton of Uri, near to the Saint-Gotthard Massif. 16 pupils of grade 5 and 23 pupils of grade 6 were taught in 4 parallel groups during 3 half-day blocks, in the morning of March 14th, 16th, and 18th 2011. Three weeks before, 4 teachers, the school director as well as a local politician were introduced to Logo within a three hours course. Each group of pupils was supervised by two tutors as well as by one class teacher. Every group acted independently from the others. One computer for each of the pupils was available. The last 90 minutes of the school project were reserved for a programming contest and for the common closing act.

The school board initiated a survey, mainly aiming to assess the impression among all the pupils and two of the teachers. Table 1 shows selected items from the survey. Each item was graded on a scale from 1 (I totally disagree) to 6 (I totally agree). The pupils and the teachers really enjoyed the project. Their feedbacks are very consistent. One teacher pointed out, that he would like to install XLogo at home, but he would need a short break in his time intensive job. Furthermore, we noted that not all the kids knew where they can download XLogo.

5.3   School Projects in Domat/Ems and in Saas im Prättigau

The Higher School of Pedagogics of the Swiss Canton of Graubünden is very interested in supporting the introduction of computer science at the local primary schools.

During two large pilot projects in 2010 and 2011, in Domat/Ems, our team was able to teach up to 120 kids of grades 5 and 6 during three half-day blocks, as well as to train their teachers. A summary of the 2010 project is available online [8]. The statements of the pupils and of the teachers are consistent with those of Attinghausen.

On May 12th, 19th, 26th and June 9th 2011 we taught a grade 5 class comprising 17 pupils at the primary school of Saas im Prättigau. The usual half-day blocks were split into two 90 minutes units just before and just after lunch. The large and comfortable classroom allowed to teach all kids together, with the supervision of two or three tutors and of the class teacher. One computer for each of the pupils was available. As usual, the last 90 minutes of the school project were reserved for a programming contest and for the common closing act.

Table 1. Selected items from the Attinghausen's survey

	grade 5	grade 6	teachers
I liked the project. I recommend it to other schools.	6 ●⊢⊣⊣⊣⊣ 1	6 ●⊢⊣⊣⊣⊣ 1	6 ●⊢⊣⊣⊣⊣ 1
The teaching materials are well-arranged and are comprehensible.	6 ⊢●⊣⊣⊣⊣ 1	6 ⊢●⊣⊣⊣⊣ 1	6 ●⊢⊣⊣⊣⊣ 1
I found it proper and important, that each pupil was allowed to individually use one computer.	6 ●⊢⊣⊣⊣⊣ 1	6 ●⊢⊣⊣⊣⊣ 1	6 ●⊢⊣⊣⊣⊣ 1
My perception of the coaching by ETH tutors and class teachers was positive. Tutors helped me by questions and supported me with hints.	6 ●⊢⊣⊣⊣⊣ 1	6 ●⊢⊣⊣⊣⊣ 1	6 ●⊢⊣⊣⊣⊣ 1
Working in four small groups, instead of the whole class, was helpful for this project.	6 ⊢●⊣⊣⊣⊣ 1	6 ●⊢⊣⊣⊣⊣ 1	6 ●⊢⊣⊣⊣⊣ 1
The closing act, with programming contest and dia-show was good and motivating.	6 ⊢●⊣⊣⊣⊣ 1	6 ⊢●⊣⊣⊣⊣ 1	6 ⊢⊣●⊣⊣⊣ 1
The distribution of the courses over 3 mornings was optimal.	6 ●⊢⊣⊣⊣⊣ 1	6 ⊢●⊣⊣⊣⊣ 1	6 ⊢⊣●⊣⊣⊣ 1
I like to occasionally program with XLogo and would enjoy to take part to a continuation of the school project.	6 ⊢●⊣⊣⊣⊣ 1	6 ⊢●⊣⊣⊣⊣ 1	6 ⊢⊣●⊣⊣⊣ 1
I plan to installed XLogo at home (or I already did it), so that I can continue to program.	6 ⊢⊣●⊣⊣⊣ 1	6 ⊢●⊣⊣⊣⊣ 1	6 ⊢⊣●⊣⊣⊣ 1

5.4 Programming Contests

The programming contests generally consist of six to nine tasks which are very similar to the exercises the pupils had to solve during the course and they are similar over all the projects. Table 2 shows the tasks of the programming contest we organized in Saas im Prättigau. The contests are neither a school examination nor an empirical test to assess the learning achievements of the pupils. They should help increasing the motivation of the children and allow to celebrate the end of the project.

We do not prepare an official ranking of the constants, but usually gift the participants solving all the tasks (or most of them) with a Logo textbook [4]. The pupils are free to choose the tasks they like to solve first and are allowed to submit their solution to a tutor several times, until they have the correct one. Therefore, we only reward correct solutions with one point. Only in very particular situations (e.g. when the time is running out and a child submits a good but not perfect solution) we distribute a fraction of a point.

Figure 1 summarizes the results for each of the tasks of the contest of Saas im Prättigau: 15 of the originally 17 pupils attended the contest. The pupils were expected to solve 7 exercises resp. 9 different tasks. The class reached an average of 6.4 out of 9 points. 60% of the children solved at least 6 tasks, 47% of them solved 7 tasks and 3 pupils were able to solve all the nine tasks. These remarkably good results confirmed the excellent feeling we had during the lectures.

The Attinghausen project consisted of only three half-day blocks. The programming contests was therefore easier and shorter. Only two tasks required to be solved using parameters. The two overall best participants were two grade 6 girls. They were able to solve all the six assignments very rapidly, within 31 resp. 33 minutes.

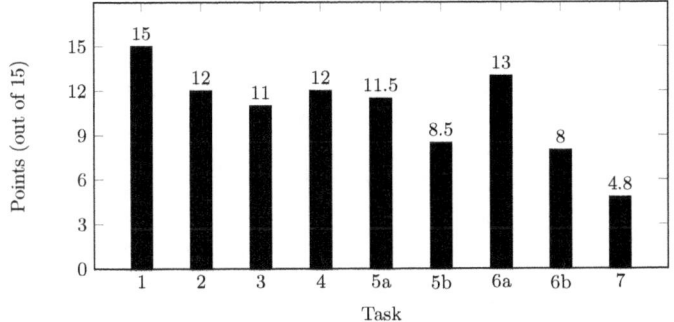

Fig. 1. Saas im Prättigau: Points per Task

6 Research Prospects

The very exiting and promising teaching experiences we made in the schools confirmed us in our conviction that extended, long-term empirical research on Computational Thinking should not be postponed. In this section, we first consider the general idea of assessing Computational Thinking, we focus on exemplary research topics dealing with the concept of a variable, and finally shortly point out our next steps.

6.1 Assessing the Influence of Computational Thinking

It can be reasonably assumed that the interdisciplinary nature of computer science may encourage a specific transfer of Computational Thinking to other scientific fields. Long-term empirical research projects are needed in order to formally investigate the impact and the probable benefits of increasing Computational Thinking skills on closely related school areas such as mathematics, natural sciences and technics. We believe that starting to program at a very early school stage plays a central role in learning Computational Thinking, and suggest to start programming classes, within a spiral curriculum, at primary schools.

Table 2. Programming Contest of Saas im Prättigau

1. Let a program draw the following picture.

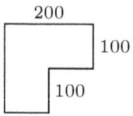

2. Draw the following picture. Use **repeat**.

3. Write a program for this figure. Use **repeat**.
The gray lines must not be drawn

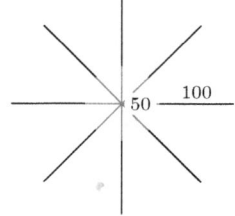

5. Fill in the necessary parts in the given programs in order to draw the picture.
a) `repeat 9 [repeat 7 [fd 40 rt 90] ...]`
b) `fd 40 rt 90 fd 360 rt 90 fd 40 rt 90 fd 40 repeat 8 [rt 90 fd 40 ...]`

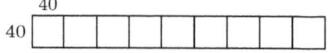

4. The program
`HALFCIRCLERIGHT`
`HALFCIRCLELEFT`
`HALFCIRCLERIGHT`
should draw the following picture. Write the subprograms `HALFCIRCLERIGHT` and `HALFCIRCLELEFT` which allow to achieve this goal.
You can freely choose the dimension of the semi-circles.

6. a) Write a program `HEXAGON :GR` to draw a regular hexagon with side length `:GR`.

b) Use `HEXAGON :GR` as a sub-program of the program `PATTERN :GR` to draw the following picture:

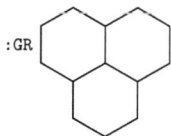

7. Write a program which draws the following picture. Users should be free to choose the side length of the squares and the number of black squares. There are as many gray squares as black squares.

In hypothetical experimental settings, a group of pupils receives an additional, excellent education in computer science, while the other groups independently get an additional, excellent education in closely related school subjects. A control group has no additional education at all. This or similar experiments, even experiments with additional education in not-MINT-topics, and the comparison of the overall learning progress of the scholar groups over several years may help addressing the influence of Computational Thinking in the context of general education.

Variations of these experimental settings allow to assess the impact of Computational Thinking skills on specific groups of pupils, e.g., female pupils, very young pupils, pupils with an high IQ or pupils with a low IQ.

6.2   Research on the Concept of a Variable

The concept of a variable is usually the first crucial issue for programmer novices of each age. Its cognitive complexity can be reduced by splitting the learning process into two phases. The pupils first learn to deal with a constant parameter, relying on a simple placeholder abstraction. The primary school kids of our school projects usually master this abstraction level. In a second step, eventually at a higher school stage, the pupils learn that parameter values may vary over time, and that a variable parameter is intrinsically related to computer memory. A simple, but adequate abstraction allows to represent it as a register. How old should children be at least, in order to master this approach, in particular the step from a constant to a variable parameter? What kind of abstraction skills or prior knowledge do they need? Is the concept of a variable parameter generally too hard for primary school kids?

Mastering variables in computer science may open new, even unexpected didactic perspectives. One of the most fascinating and challenging research prospect refers to the only apparently similar role of variables in computer science and mathematics. Children starting to program early, in the primary school, do not have any school related prior knowledge on variables. They therefore develop this concept from scratch. Even relying on the register abstraction, a variable in computer science is more concrete than in mathematics. Children who first learn to program and later have to deal with the more abstract concept of variables in mathematics classes, may benefit from their programming skills.

6.3   Upcoming Research Activities

Gaining empirical evidence of learning progress requires a sound test design as well as a large population of pupils to be followed over a long period. The pupils have eventually to be followed over most of their mandatory school, and even longer. We are looking ahead for planning and soon starting long-term research projects with partner institutions and an interdisciplinary group of scientists and teachers. Highly interested primary schools are already available. Classes at higher school stages have to be found.

7   Conclusion

This paper considers Computational Thinking as the key contribution of computer science to general education. It presents our experiences in introducing primary school kids to Computational Thinking by teaching them how to program. In our projects, primary school kids are expected to learn autonomously, without a systematic direct instruction. The pupils solve small algorithmic problems by programming in Logo, relying on very few commands in a simple editor. They are therefore much less exposed to the undesirable effects of cognitive overload.

Even relying on an appropriate sound didactic concept, programming remains a very challenging cognitive activity. Modular design is one method the kids learn in order to address this complexity. Observation during the courses, the results of the programming contests, and the feedback of pupils and teachers informally confirm that the kids achieve the goals we set, and point out the quality of the chosen didactic approach, independent from the gender of the pupils.

Computational thinking may permit specific transfer from computer science to the other MINT subjects, and may eventually open new didactic opportunities. Extensive research projects are needed to gain empirical evidence of the its importance.

References

1. Brusilovsky, P., Calabrese, E., Hvorecky, J., Kouchnirenko, A., Miller, P.: Mini-languages: a way to learn programming principles. Education and Information Technologies 2, 65–83 (1998)
2. Freiermuth, K., Hromkovič, J., Steffen, B.: Creating and testing textbooks for secondary schools. In: Mittermeir, R.T., Sysło, M.M. (eds.) ISSEP 2008. LNCS, vol. 5090, pp. 216–228. Springer, Heidelberg (2008)
3. Hromkovič, J.: Contributing to General Education by Teaching Informatics. In: Mittermeir, R.T. (ed.) ISSEP 2006. LNCS, vol. 4226, pp. 25–37. Springer, Heidelberg (2006)
4. Hromkovič, J.: Einführung in die Programmierung mit Logo. Vieweg+Teubner (2010)
5. Hromkovič, J.: Informatik und allgemeine Bildung (May 2010),
   http://www.educ.ethz.ch/unt/um/inf/all_inf/unt/um/inf/all_inf/
6. Hromkovič, J., Keller, L., Serafini, G., Steffen, B.: Programmieren mit Logo,
   http://abz.inf.ethz.ch/primarschule-unterrichtmaterialien
7. Le Coq, L.: Xlogo. Website, http://xlogo.tuxfamily.org/
8. Matter, B.: Projekt programmieren in der primarschule. Website,
   http://abz.inf.ethz.ch/media/archive1/programmierenfuerkinder
   /InfobroschAug2010-3.pdf
9. Papert, S.: Mindstorms: Children, Computers and Powerful Ideas, 2nd edn. Basic Books, New York (1993)
10. Schnabel, R.B.: Educating computing's next generation. Commun. ACM 54, 5 (2011)
11. Wing, J.M.: Computational thinking. Commun. ACM 49(3), 33–35 (2006)

Learning Algorithmic Thinking with Tangible Objects Eases Transition to Computer Programming

Gerald Futschek and Julia Moschitz

Vienna University of Technology,
Institute of Software Engineering & Interactive Systems,
Karlsplatz 13, 1040 Vienna, Austria
{Gerald.Futschek,Julia.Moschitz}@ifs.tuwien.ac.at

Abstract. Learning algorithmic thinking can start in early years and must be oriented on the thinking ability of young children. Suitable environments with tangible objects and easy to understand problems motivate the young to learn the first concepts of algorithms. We present in this paper a learning scenario Tim the Train for primary school children, that involves tangible objects and allows a variety of interesting tasks to learn basic concepts of algorithmic thinking. We also show how a smooth transition from a playful environment with tangible objects to a virtual Scratch/BYOB environment may help the young learners to learn their first steps in understanding virtual environments and programming concepts.

Keywords: Algorithmic thinking, primary education, explorative learning, tangible objects, learning by doing.

1 Introduction

At the beginning of learning algorithmic thinking learners often have problems with this abstract type of thinking and with the abstraction in a programming environment. Therefore, we try to build a bridge between the real world and a programming environment. *Tim the Train* is an example which helps especially beginners to pass the river between the tangible world and the abstract world of programming environments and languages.

Many efforts to create environments and learning scenarios have been made in order to make learning algorithmic thinking and programming easier, more pleasant and more efficient. Some more prominent programming environments that are suitable also for very young children are Logo, Alice, Baltie and Scratch [1,2,3,4]. All these environments allow the manipulation of virtual objects with a wide variety of commands. Learning environments should try to attract as many learners as possible by appealing to different senses and learning styles. For young learners the kinesthetic aspect of learning is very important, because they learn by touching, seeing, hearing, feeling and smelling [5]. Kinaesthetic elements are very important even for engineering education in higher education, because the students are invited to be active with more than one sense in the learning process [6]. Some programming environments involve tangible objects, the mechanical Logo turtle and programmable robots are well-known examples [7].

I. Kalaš and R.T. Mittermeir (Eds.): ISSEP 2011, LNCS 7013, pp. 155–164, 2011.

In this paper we propose a learning scenario with real tangible objects that is suitable to learn the basic concepts of algorithmic thinking. It has a limited number of commands that are easy to understand. A simulation of this scenario in Scratch allows a smooth transition to virtual learning environments and the world of computer programming.

2 Learning Scenario *Tim the Train*

Tim the Train describes a bin packing scenario. The containers of a train should be loaded with parts of different shapes. No container should be overloaded. This learning scenario provides the opportunity to come in contact with a more advanced problem of computer science. *Tim the Train* helps beginners to train algorithmic thinking in a playful way and to understand some aspects of the bin packing problem as well.

The implementation of *Tim the Train* consists of a set of wooden elements for train, containers and load parts (see Fig. 1) and also of a simulation in Scratch/BYOB. *Tim the Train* includes a locomotive, containers, parts in different shapes and sizes and a set of symbolic commands. The colours of the train are chosen so that they are attractive for boys and girls. The wooden train and the simulation in Scratch/BYOB use the same structure and colours, so that a recognition factor is given, which helps beginners to get a feeling for the simulation environment on the computer.

Fig. 1. *Tim the Train* with tangible parts made of coloured woodIntroduction

A variety of interesting tasks can be performed with this learning scenario. Basically, learners do not need special pre-knowledge to solve these assignments. The game-like tasks are suitable for children from the age of 5 years onwards. Reading is not necessarily required, because the instructions for the algorithms are marked with symbols and words. The teacher should explain the tasks and the meaning of the commands anyway. The number of symbols for describing the commands (a sort of programming language) is kept small to help the learners to focus on the basic algorithmic concepts. The learners get a card with explanations about the actions and their acceptation. The programming language is easily expandable, because the language is realized on cards. New command cards may be created by the learners themselves. For advanced learners it is a challenge to create new actions or forms of loops and suitable symbols for these actions and loops.

The problems (tasks) are chosen in a way that they are interesting for beginners as well as for advanced learners, because the tasks provide different levels of difficulties. Advanced learners can solve the easier problems too in order to revise and consolidate the basic skills of algorithmic thinking. However, the material is also appropriate to adolescents and adults who want to learn algorithmic thinking in a playful way.

3 Basic Algorithmic Concepts

Learning concepts is much more important than learning systems or programming languages. At the very beginning of learning algorithmic thinking we can observe very basic concepts that should be involved in learning scenarios:

> Basic commands – basic actions
> Sequence of commands
> Alternative of commands (if)
> Iteration of commands (loop)
> Abstraction command (method)

Commands are usually represented by symbols or by text. The main purpose of a command is that when executed it invokes well defined actions. Each basic command is related to a basic action in the learning scenario. So a learner has to understand that a command is an abstraction of an action. The concepts sequence, alternative and iteration define the order in which commands are executed. These concepts are essential in algorithmic thinking.

Other important algorithmic concepts like recursion, parameters, variables, data types are omitted intentionally when addressing very young learners. These concepts need deeper abstract thinking skills and are therefore suitable only for elder learners or more advanced learners.

4 Interesting Tasks

This chapter describes three different tasks for beginners to play/learn with *Tim the Train* that lead to first algorithmic thinking skills. The level of difficulty increases from task to task. These three tasks are just examples of tasks that can be given. A teacher may vary these tasks or can pose even his own tasks.

4.1 Playing a Given Algorithm

In the first level the beginners have to reproduce a given algorithm. By this way the beginners familiarize themselves with the symbols (commands) of *Tim the Train*. Additionally, the learners try to play (execute) the given algorithm and to find out the result of the given algorithm. This game is appropriate for beginners who have no or only few experiences of algorithmic thinking. This problem is also adequate for very young children, who cannot read or write. At the beginning the containers of the train (Fig. 2) contain some parts to show what the action "reset" does at the beginning.

Fig. 2. A train at the beginning of the algorithm

The following table shows the parts and the order of their appearance.

Table 1. A sequence of incoming parts

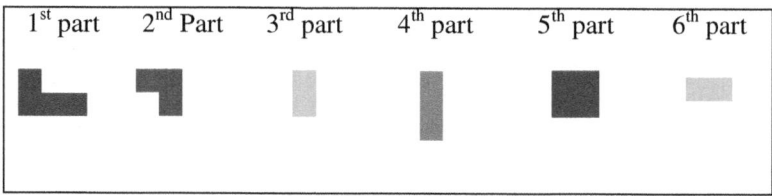

1st part	2nd Part	3rd part	4th part	5th part	6th part

The first command (reset) of the algorithm (Fig. 3) is to clear all containers of the train. Then the train is empty and the first part can be dropped. The instruction "drop" means to drop the part at the leftmost position where the part would reach the deepest place. This step will be repeated four times. After these drops the last part overloads the wagon therefore it is lifted again. Then the next wagon is loaded with three parts.

Fig. 3. Example of a given algorithm

The learners have two possibilities to build the solution. On the one hand they can play the algorithm with the wooden train, on the other hand they can even use the simulation in Scratch/BYOB to find the solution. When the algorithm is finished, one can find the following final state (Fig. 4):

Fig. 4. The train at the end of the algorithm

4.2 Second Task: Writing an algorithm

In the second task, the result (Fig. 5) is given and learners have to find out the algorithm that leads to the given result.

Fig. 5. The final state to be achieved

For beginners a set of instructions cards (Fig. 6), which are required in the algorithm, may be given.

Fig. 6. A given set of instruction cards

As additional help to find out the basics of the algorithm, a few tips may help: "At first you have to find out in which order the parts are fitted in the wagons.", "Check the parts, if they are in the right position. Maybe you have to flip or rotate some of the parts."

Advanced learners may begin to solve the problem without the tips and the set of given commands. As an additional task, advanced learners may try to find a better solution by use of iterations.

If the learners use the actions shown in Fig. 6, the parts have to come in the following order and orientation:

Table 2. The sequence of incoming parts

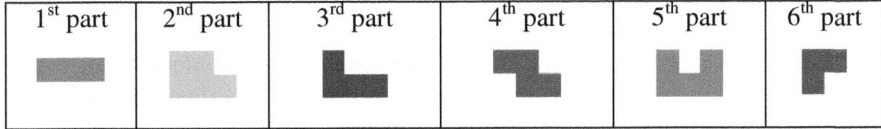

1st part	2nd part	3rd part	4th part	5th part	6th part

The solution with the instructions given is without any use of loops. Therefore, it will be easy to solve for beginners (Fig. 7). The beginners can think about: What is happening when the parts are in another order or orientation? How must the algorithm be changed to reach the given result? Does the algorithm always work for all possible input sequence of parts? These questions are instructive for beginners, because they will recognize by this way that the solution is only effective for the given sequence of parts and not a universally valid solution for this kind of problem.

Fig. 7. A possible solution for the required algorithm

4.3 First Algorithm with Loops

The next task allows beginners to develop repeat–loops in an easy way. The didactic approach is that the learners develop the loop step by step. At first the learners have to find out what a given algorithm does and then they have to reformulate the given algorithm using the loop construct. Next they have to find out: "Under what condition is the algorithm correct and under what condition not?" This question should help especially beginners to think about the differences between an algorithm which is generally effective and one that is effective only for a specific problem. The precondition for the algorithm of Fig. 8 is that the train has at least three wagons and overloading is possible. As additional difficulty, the learners do not know in which order the parts are coming. The following algorithm drops a part in each of the first three wagons. This action is repeated three times.

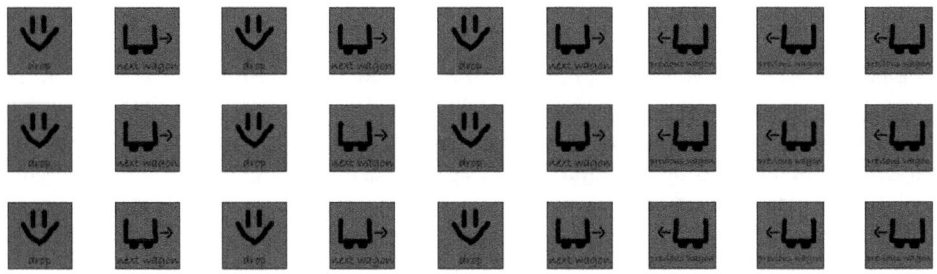

Fig. 8. A given simple algorithm at the beginning of the task

At first the learners have to find out which parts of the algorithm are repeated. It is easy to see that the sequence of commands contained in the 3 rows is exactly the same. But also inside each row one can find repetitions. The left block in Fig. 9 shows the actions "drop and go to next wagon" which are repeated 3 times. The right block in Fig. 9 shows that the action "go to the previous wagon" is also repeated three times.

Fig. 9. The repeated parts of each row are marked by coloured blocks

As intermediate step the learners may build an algorithm using loops only for the two blocks in Fig.9, because loops, especially nested loops, are not so easy to understand. For beginners the breaking points of loop understanding are to find the repeated actions and to set the right limitation of the loop.

An example of a solution is given in the following picture (Fig. 10). This solution shows explicitly, that the two loops of the first line are repeated three times.

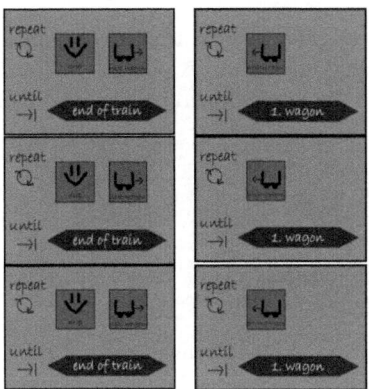

Fig. 10. An algorithm with six loops

Additionally, by this way the learners discover that they can simplify the solution once again and develop their first nested loop in an easy way. A solution with an additional loop may simplify this algorithm even further, see (Fig. 11).

Fig. 11. A solution with nested loops

This example shows how loops can be introduced to beginners. Step by step the learners develop new versions of algorithms.

As additional tasks it would be interesting to write algorithms to fill a wagon and to empty a wagon. These loops are easy to build, but they are good exercises to develop skills in loop understanding. The algorithm to empty a wagon is simple, because one needs only the action "lift" and the condition "empty" (Fig. 12).

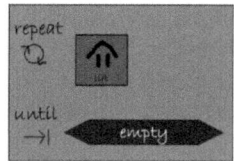

Fig. 12. Emptying a wagon

Filling a wagon is only slightly more difficult. The additional difficulty is to detect that a wagon cannot be filled any more. A possible solution is to drop parts until the wagon is overloaded. Then the last dropped part has to be lifted back (Fig. 13).

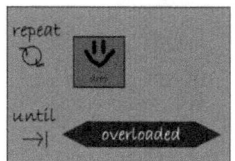

Fig. 13. Filling a wagon

5 Learning Experiences

All the examples described in the last chapter try to help beginners to train their competences in algorithmic thinking. Although the complexity of *Tim the Train* is simple, it has a lot of possibilities to construct interesting tasks for different levels of algorithmic thinking skills.

Fig. 14. Filling the wagons with tangible parts

For learners who have problems with algorithmic thinking or who are using preferably visual or kinaesthetic learning styles, the wooden train helps them to understand the problem and eases finding a solution (see Fig. 14). This type of learners is grateful, when they can touch and flip the real parts in the real world or to execute an algorithm without a computer environment. At the beginning these learners are often overstrained with doing both: algorithmic thinking and at the same time orienting in a new abstract programming environment.

We strongly recommend that the learners work in groups. With help of the wooden train the learners can work together, can discuss their problems, and they can find solutions to their problems faster.

In the examples described *Tim the Train* is standing and a crane is moving from one container to another. Advanced and more talented learners like to find new tasks or to expand the set of commands. For example, it would be interesting to move the crane or the train in both directions. This learning scenario would enable a lot of further solutions and a basis for discussions of the different solutions.

Advanced learners are highly motivated to implement their commands also in Scratch or BYOB [8]. The principles of algorithmic thinking are in a virtual programming environment like Scratch or BYOB the same as with tangible objects. It differs in the granularity of the basic actions. Even the basic commands of the wooden model have to be programmed, see Fig. 15.

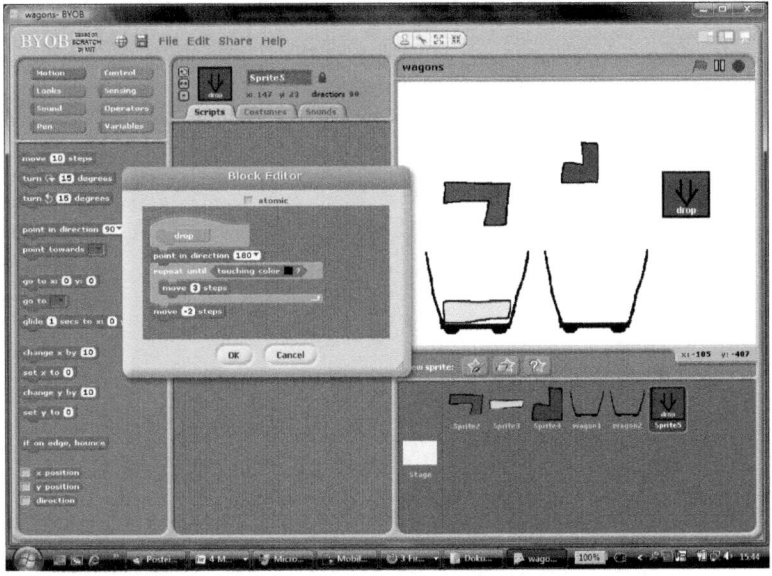

Fig. 15. Implementation of command "drop" in BYOB

The advantage of this learning scenario is that the learners may decide between different learning environments adapted to their individual learning styles and types.

6   Conclusions

Tim the Train is a physical micro world which helps learners to train algorithmic thinking and helps them to have a smooth transition to virtual learning environments and the world of computer programming. The children learn about the basics of the bin packing problem and make their first steps in the programming environment of Scratch.

With *Tim the Train* the beginners are motivated to do further programming in different environments. This holds even for those learners who are not absolutely ecstatic of programming at the beginning.

References

1. Papert, S.: Mindstorms: Children, computers, and powerful ideas (1980)
2. Cooper, S., Dann, W., Pausch, R.: Teaching Objects-first. In: Proceedings of the 34th SIGCSE Technical Symposium on Computer Science Education, Introductory Computer Science, pp. 191–195 (2003)
3. Baltie home page, http://progopedia.com/language/baltie/
4. Maloney, J., Burd, L., Kafai, Y., Rusk, N., Silverman, B., Resnick, M.: Scratch: A Sneak Preview. In: Second International Conference on Creating, Connecting, and Collaborating through Computing, Kyoto, Japan, pp. 104–109 (2004)
5. Hayes, D.: Encyclopedia of Primary School, pp. 231–232. Routledge, Chapman & Hall (2010)
6. Felder, R., Silverman, B.: Learning and Teaching Styles in Engineering Education. Journal of Engineering Education 78(7), 674–681 (1988)
7. Horn, M., Jacob, R.: Designing Tangible Programming Languages for Classroom Use. In: 1st International Conference on Tangible and Embedded Interaction, pp. 159–162. ACM, New York (2007)
8. BYOB - Build Your Own Blocks, An extension of Scratch to program own commands, http://byob.berkeley.edu/

Transfer, Cognitive Load, and Program Design Difficulties

David Ginat[1], Eyal Shifroni[2], and Eti Menashe[1]

[1] Tel-Aviv University,
Science Education Department,
Ramat Aviv, Tel-Aviv, Israel 699978
ginat@post.tau.ac.il, etime@zahav.net.il
[2] Tel-Hai College,
Computer Science Department
eyalshif@telhai.ac.il

Abstract. We display a series of five studies of student difficulties with transfer during the design of computer programs. The difficulties are characterized with five transfer aspects – recognition, abstraction, mapping, embedment, and flexibility. Each study involves a programming task, and unfolds difficulties with one or more of the above aspects. The majority of the posed tasks were rather simple CS1 (Computer Science 1) or CS2 tasks, and involved specific transfer. One of the posed tasks was more involved and required both specific and non-specific transfer (and a subtler combination of the above aspects). We tie our findings to the notion of cognitive load, and its sub-notion of intrinsic cognitive load. Following our findings, we offer recommendations and guidelines for tutors, for developing improved transfer in program design.

1 Introduction

"*Transfer* occurs when a person's prior experience and knowledge affect learning or problem solving in a new situation" ([5], p. 48). In the domain of programming, we regard *transfer* as the effect of one's prior design experience and knowledge in new program design situations. In particular, transfer may appear as the invocation and utilization of suitable programming patterns [1,8], as well as other design elements, such as top-down decomposition or modelling.

Transfer involves a few aspects, among them: recognition, abstraction, and mapping [5]. *Recognition*, in the sense of a reminded, potential analogue (such as an analogous problem, scheme, or representation); *abstraction*, in the sense of an identified general structure, which may be diversely manipulated; and *mapping*, in the sense of successfully relating the recognized analogue to the task at hand. Diverse manipulations and mapping involve flexibility and embedment – *flexibility* in the sense of "local" modifications, in order to adapt to the task at hand; and *embedment* in the sense of combining one element with another, such that the second element "serves as a conduit" to the first [4].

The notion of *cognitive load* refers to the number of cognitive elements (possibly patterns, or resources), which are handled during learning, or problem solving, in

I. Kalaš and R.T. Mittermeir (Eds.): ISSEP 2011, LNCS 7013, pp. 165–176, 2011.

one's working (short term) memory [4,9,10]. Yet, not only the number of cognitive elements is relevant, but also the elements' interactivity. The aspect directly related to element interactivity is considered as *intrinsic cognitive load*, as demands on working memory capacity, which is imposed by element interactivity, is regarded as implicit to the learned elements [7].

Cognitive load in general, and intrinsic cognitive load in particular, are considered to be directly related to knowledge organization in one's long-term memory. Well-organized long-term memory involves structured and richly connected cognitive schemes, encapsulating one's knowledge and skills. Such organization reduces cognitive load upon learning and problem solving, and yields higher competence. One of the primary differences between experts and novices derives from their different knowledge organizations, e.g., [3,10].

A variety of research studies in the last two decades advocated the development of students' knowledge organization, by explicit presentation of basic programming, or algorithmic patterns (also called design patterns, templates, schemes, and idioms [1]). These patterns were specified by experts, primarily due to their repeated, or reusable occurrences as code building blocks. Some fundamental algorithmic patterns are: counting, summation, max-computation, and previous-current progression (as in the computation of the Fibonacci sequence).

Suitable utilization of algorithmic patterns involves the transfer aspects indicated above, of recognition, abstraction, mapping, flexibility, and embedment. These, in turn, are related to the amount of cognitive load (including intrinsic cognitive load) upon problem solving.

In particular, upon solving an algorithmic/programming task, one should recognize the relevance of particular patterns, possibly abstract them, and/or use them with some flexibility, and/or embed them together, and map the outcome to the task at hand. Well-organized, structured conception of algorithmic patterns, which also includes their interconnections, may yield reduced cognitive load upon problem solving, enable suitable transfer, and improve competence.

Studies such as [3] offered structured frameworks for algorithmic patterns' organization. Yet, little was studied about concrete student difficulties with algorithmic patterns' employment. Two exceptions are [6,8], which examined pattern composition, in terms of goals of combined sub-components. Yet, all of these studies did not examine difficulties with design transfer, and the aspects of recognition, abstraction, mapping, flexibility, and embedment. A study of these aspects, with relation to the notion of cognitive load, may help tutors better understand their student behaviours, and may yield improved teaching practices.

In this paper, we display a series of studies that illuminate a variety of difficulties with respect to aspects of transfer during program design. We reveal the difficulties through five examples with CS1/CS2 high-school students, high-school teachers, and college students. In the next section we present student difficulties and relate them to elements of transfer and cognitive load. In the section that follows, we discuss our findings, relate them to implications on teaching, and offer some recommendations and guidelines for CS tutors.

2   Design Transfer Difficulties

We display below five studies, which we conducted with high-school students, high-school teachers, and college students, in order to examine the transfer aspects indicated in the Introduction. Each of the following five subsections presents one study. Each study involves one programming/algorithmic task, of sequence processing, which we posed to the participants. The task in the first study was elementary, and the tasks in the studies that follow were, gradually, more involved. None of the tasks were challenging, apart from the last one. Yet, the participants demonstrated various difficulties, which differed between the tasks. In each subsection, we first indicate the population and display the task. Then, we describe the participants' solutions, and analyze them with respect to transfer aspects and the notion of cognitive load. We name each subsection according to the dominant transfer difficulty (or difficulties) revealed by the task posed in its study.

2.1   Difficulties with Flexibility

The first population included 95 11th grade, high-school students, who have completed their first of two CS1 years (which they studied with JAVA). They learnt diverse utilizations of *if*-statements, loops, and arrays, with elementary algorithmic patterns, such as counting, summation, max-computation, and integer decomposition (into digits). We posed the following elementary task. Although very elementary, we still (surprisingly) noticed difficulties.

> **Last rainy day.** Develop a program for which the input is 365 integers indicating the amount of rain in each day of the year; and the output is the (index of the) last rainy day.

Trivial? Indeed; so it was for the vast majority of the students; but not for all of them. The solution of this task is, in a sense, a variant of the max computation, in which every input element is compared to some value, and a corresponding variable is updated accordingly. In the max computation, the comparison is to the value of the *max* variable; here, the comparison is to the constant 0. If greater than 0, a *last* variable should be updated, according to the location (index) of the last-read element. The solution may also be seen as a variant of conditional counting in which each read element is checked, and a counter is updated accordingly. Here, there is no counter, but rather a location variable.

Max and counting computations are basic, "canonical" algorithmic patterns, with which all the students were familiar. In solving this task, students only needed to flexibly utilize one of these patterns in order to yield the necessary solution.

Still, 26 students (27%) provided an erroneous solution. Some may have misread the task (e.g., computed the last non-rainy day); but others demonstrated a difficulty in devising a suitable solution. In particular, 12 students (13%) simply applied one of the above "canonical" patterns (max, counting), or offered an unsuitable variant of these patterns. Several interviews with these students unfolded a vague adaptation of the invoked "canonical" patterns. The students "felt" that they had to offer a structure similar to their invoked patterns, but "did not exactly manage to do so".

Viewing the above phenomena in transfer "lenses", we notice that the task's solution required transfer that involved *flexibility*; in the sense of simple, flexible manipulation of a selected, basic pattern. The 12 indicated students were familiar with the relevant "canonical" patterns, and have previously seen various variants of these patterns. They had to recognize the relevance of any of these patterns, select one, flexibly modify it, and map it to the given task. They did perform the recognition part (of the relevant pattern), but failed to properly perform the flexible modification that maps the recognized "canonical" pattern to the task at hand.

In our impression, the above phenomenon may also be related to cognitive load effects. The students who demonstrated difficulties seemed to struggle with simple variations of "canonical" patterns. They tried to manipulate the relevant patterns (which they properly recognized), but their intrinsic cognitive load, which derived from their limited view of pattern interactivity, and pattern interconnections, yielded what seemed to be a "blurred" picture of possible element (pattern) manipulations.

2.2   Difficulties with Recognition

The second population included 45 participants, composed of 12 college CS2 students and 33 high-school teachers, who had different backgrounds. Some of these teachers completed ordinary undergraduate CS studies, while others majored in different scientific domains, but learnt the CS1 and CS2 courses. All the participants studied and practiced diverse, simple and complex utilizations of fundamental algorithmic patterns. Nevertheless, some demonstrated difficulties with the following basic task.

> **Number of subsequences.** Develop a program for which the input is a long sequence of 1's and 0's, starting with 1 and ending with 1; and the output is the number of subsequences of 0's.
> Example: For the input 1 1 0 0 1 0 1 0 0 0 1 1 1 0 1, the output should be 4 (we consider a subsequence composed of one 0 as a valid subsequence).

Another trivial task. So it was for some of the students, but again, only for some. An elegant solution of this task is based on the observation that the number of desired subsequences is equivalent to the number of starting-points of such subsequences, or the number of endpoints of such subsequences. Thus, it is sufficient to count, for example, the number of endpoints of subsequences of 0's. Each endpoint is characterized by a 0 that is followed by a 1. In devising the solution, one may compose the two basic algorithmic patterns of previous-current and counting. The former pattern will be used for identifying the endpoint of a subsequence, and the latter – for counting the number of such endpoints. The two patterns should be composed in an interleaved way, which encapsulates an on-the-fly computation.

In order to solve the task, one had to recognize the observation described above, and then recognize that the solution is an interleaved composition of the previous-current pattern with the pattern of counting. This interleaved composition is actually an embedment of the former pattern in the latter one.

Only 19 participants (42%) yielded the described elegant solution. The rest of the participants demonstrated recognition and embedment difficulties. 10 students (22%) did recognize the above observation (of starting-points/endpoints), but offered a rather cumbersome solution, which involved a flag. They used a flag in order to indicate a state of "currently reading 1's". When the flag was "on" and the next input element was 0, the subsequences counter was increased by 1. The idea underlying this solution is correct, but its implementation is error-prone. And indeed, some of these participants yielded erroneous solutions, particularly with handling the flag. It seemed that these participants did not recognize the relevance of the algorithmic pattern of previous-current for solving the task.

The rest of the participants demonstrated more profound difficulties. 9 of them (18%) did recognize that the subsequences counter may be increased once a 0 appears after a 1, but were unable to scan the input in a suitable way. Their code included a section for reading a subsequence of 0's, without properly handling the end of such a subsequence. Some of their solutions included this section in a separate procedure, whereas others involved nested loops that did not properly terminate once the end of a subsequence of 0's was reached. Thus, they demonstrated a deeper difficulty with recognizing the suitable pattern to perform the computation.

The remaining (8) participants (18%) offered cumbersome solutions that did not reveal any insight into the task. In particular, they did not demonstrate any recognition of a suitable way for conducting the necessary counting (some counted the total number of 0's; others initialized the counter time and again).

Viewing the above phenomena in a transfer perspective, we notice that the task solution required two kinds of the transfer aspect of *recognition* – the recognition of the task characteristic (of starting-points/endpoints), and the recognition and selection of suitable algorithmic patterns. More than half of the students demonstrated difficulties with the second kind, and about a fifth demonstrated difficulties already with the first kind.

Here again, the observed difficulties seemed to be related to cognitive overload, which "blurred" the picture of what to focus-on in the input, and how to "translate" an insightful task observation into a suitable solution scheme. In particular, the participants who tried to design a solution which "collects" the 0's of each subsequence felt confused and uncertain in devising a scheme that properly terminates such a collection, and seemed to have a vague picture of how to adapt their chosen pattern to the desired solution scheme.

2.3 Difficulties with Embedment and Abstraction

The third population included 31 college students, who have completed the CS1 course (in JAVA), and were in the end of their CS2 course. We posed them a rather more involved programming task.

> **Longest stair.** We define a stair as a plateau of integers (a sequence of one or more equal integers) followed by another plateau of integers that are greater by 1 (e.g., 1 1 2 2 2). Develop a program for which the input is a very long sequence of integers, such that each integer is equal-to or greater-by-1 than the previous one; and the output is the length (i.e., the number of integers) of the longest stair. (Notice that stairs overlap.)
> Example: For the input 2 2 3 3 4 4 4 5 6 6 6 7, the output should be 5 (due to the stair 3 3 4 4 4).

This task is more involved than the previous two tasks, yet its solution is still relatively short. It is based on an on-the-fly computation, of 'one pass' over the input (notice that the input is very long), and involves an interleaved composition of four algorithmic patterns:

- A pattern of previous-current, for comparing the last two read integers.
- A counting pattern, for keeping the location of the last-read integer.
- A max pattern, for keeping the length of the currently longest stair.
- Another pattern of previous-current, for keeping the starting points of the last two plateaus that compose the latest stair.

In the on-the-fly computation, the length of the latest stair (the last two plateaus), may be computed when the first integer of a new plateau is just read. That is, the event of starting a new plateau also yields the end of the latest stair. The length of the latest stair may be obtained from adding the lengths of the just-ended plateau and the previous plateau. The computation may be performed by subtracting the starting point of the previous plateau from the location of the just-read integer. Two additional updates should be added: an update of the starting points of the two latest plateaus, and an update of the currently longest (max) stair.

The design of this on-the-fly computation requires interleaved embedding of the above four patterns. One has to first recognize the relevance of these patterns, and then properly assemble them together. The recognition and embedding of the first three patterns is rather simple, whereas the recognition and embedding of the fourth pattern is more delicate. In particular, the recognition of the fourth pattern involves abstraction of the basic previous-current pattern. (Notice that this pattern is also used in its basic form, as indicated in the first of the four algorithmic patterns listed above.)

42% of the students (13 out of the 31) constructed solutions, which combined the first three patterns, in diverse ways, but did not include the fourth pattern. Their solutions looked cumbersome and awkward. It seemed that they just assembled pattern pieces, possibly in order to demonstrate that they "can write something" (as one of them indicated in a following interview), and show that they are acquainted with comparisons of adjacent elements and some composition of basic patterns. They knew that their solutions involved only partial pieces, but could not do better.

35% of the students (11 out of the 31) properly embedded all the four patterns, but over-simplified the fourth pattern. They ignored the overlapping of stairs, and computed only every second stair. For example, for the input 2 2 3 3 4 4 4 5 6 6 6 7, they computed only the first, third, and fifth stairs: 2 2 3 3; 4 4 4 5; and 6 6 6 7. In interviews that followed, they expressed a difficulty to abstract and properly specify the fourth pattern, in a form that handles stair overlaps.

Only 22% (7 out of the 31) of the students offered suitable solutions. Some of these students assembled solutions that properly interleaved all the above four patterns, while others slightly changed the fourth pattern, and kept counters for computing plateau lengths, instead of plateau starting points.

Viewing the student solutions in a transfer perspective, we may notice two primary transfer difficulties. 24 out of the 31 students (77%) demonstrated incompetence in pattern *embedment*. In addition, 13 of these 24 students did not *abstract* the basic previous-current pattern, into a previous-current scheme of adjacent plateaus.

The other 11 students demonstrated limited abstraction, while oversimplifying the computation of adjacent plateaus.

The task solution required both pattern abstraction and the interleaved composition of four algorithmic patterns. The majority of the students, who did not yield the suitable solution, seemed to (also) be affected by intrinsic cognitive overload. Their difficulties seemed to stem from a limited picture of pattern interactivity, particularly with respect to the necessary embedment. Some students managed to handle the interleaved composition of three patterns, but felt short in doing so with four patterns.

2.4 Difficulties with Mapping

The fourth population included 33 high-school teachers, some of whom completed ordinary undergraduate CS studies, and others who majored in different scientific domains (the same teacher population of section 2.2). We posed them the following task, whose solution scheme is simpler than that of the task in the previous section, but its required insight may be slightly subtler.

Largest drop. We define the largest drop in a sequence of integers as the maximal difference between two sequence elements, such that the larger among them appears first in the sequence. Develop a program for which the input is a very long sequence of integers; and the output is the value of the largest drop
Example: For the input 8 2 3 1 10 4 4 5 6 2 7, the output should be 8 (due to the difference between the 10 and the second 2).

Upon reading the task, one clearly sees that it involves a max computation. The delicate point to observe is the relevant drops to examine. Each difference between two input elements is a (positive or negative) drop, but not all drops should be examined. Upon reading the input, the only drop that is relevant to examine after reading the next input element, is the difference between the max so-far and the newly read element. No other difference between a value other than the latest max and the newly-read element is relevant. Thus, the solution involves an on-the-fly computation, which is composed of an interleaved composition (embedment) of two max patterns – one for updating the current max, and one for updating the largest drop seen so far.

A third (11) of the participants did not notice the above observation, and designed the very inefficient solution of examining every possible drop; that is, examining the difference between every pair of elements. Their non-on-the-fly solution involved a composition of the max pattern (for the largest drop) with the pattern of thorough enumeration of all the possible pairs of elements. These students did recognize the obvious relevance of the max pattern, but fell short in elegantly mapping the utilization of this pattern into a two max patterns composition for the task at hand.

Ten other participants (30%) attempted an on-the-fly solution, but offered erroneous solutions. Some calculated the largest difference between the max of the sequence and the elements that follow it; others also considered the largest difference between the min of the sequence and the elements that precede it; and some offered some other variants of these erroneous ideas. Thus, these participants also fell short in properly mapping the utilization of the max pattern into a two max patterns scheme for the task at hand. The rest (12) of the participants (36%) did yield the suitable on-the-fly solution, of the two max patterns scheme described above.

Viewing the participant solutions in a transfer perspective, we notice that the participants did recognize the relevance of the max pattern, but about two thirds of them were unable to elegantly *map* its two-fold utilization into the task at hand. The cognitive load aspect that was apparent here was related to the participants' limited ability to recognize the intrinsic characteristics of the max pattern, and the diverse ways by which it can be employed, including the interleaved composition of two of its variants.

2.5 Difficulties with Recognition, Abstraction, and Mapping (Combined)

Our fifth example involves a rather subtle task, which we posed in a national algorithmic competition. The task was posed to 143 high-school students. Most of them completed both the CS1 and CS2 courses, and a few only completed CS1. The algorithmic pattern required for the solution was basic, but the recognition, abstraction, and mapping of this pattern into the task at hand was not trivial.

> **Kangaroo hops.** Develop a program for which the input is a positive integer d, that specifies the fixed length of a kangaroo hop, followed by a very long sequence of N increasing integers that specify the locations of barriers; and the output is a message indicating whether there is an integer location before (smaller than) the first barrier, from which the kangaroo can start jumping, and perform N jumps forward, so that it lands in each jump between (but not on any of) the next two barriers, and reach a final point past the last barrier.
> Example1: For the input 5 2 8 12 18, the output should be "Yes", since the kangaroo, whose hop is 5, may start jumping from any of the locations -1, 0, or 1, and perform successfully the desired sequence of jumps (e.g., if it will start from location 0, then it will land in locations 5, 10, 15, 20).
> Example2: For the input 5 6 8 12 18, the output should be "No", as no starting point may yield a successful sequence of more than one jump.

One way of approaching the task is by trying as starting point each of the d-1 integer locations just prior to the first barrier. For each such starting point, a sequence of N jumps will be simulated and checked, using a variant of the linear search pattern. However, this solution is not really relevant for a long sequence of barriers, as it requires keeping all the barrier locations in memory.

The suitable solution is based on recognizing the observation that we may "skip forward" with a *continuous range* of values rather than with a single integer. That is, we may start with the range of all the d-1 potential starting points before the first barrier, jump forward, and land between the first and the second barriers. This may require some reduction of the range with which we skip, from the lower-end of that range and from the upper-end. Then, we will jump again, with that remaining continuous range, and land between the second and the third barrier, and again possibly reduce the remaining range; and so on. If we will successfully perform N jumps, and reach the zone past the last barrier with a non-empty range, then the output will be "Yes". Otherwise, it will be "No".

The above solution idea requires: 1. recognition of the relevance of skipping with a continuous range; 2. abstraction of the linear search pattern of progressing with a range, rather than with a single integer; and 3. suitable mapping of that pattern into the task at hand, by carefully calculating the range reductions after every jump.

Indeed, the need to recognize and combine the latter three elements was difficult to the vast majority of the students. Nevertheless, as the task was posed in an algorithmic competition, we expected that only a small minority will provide the suitable solution.

Over one third of the students did not offer any solution. Another third offered the inefficient solution, of separately checking validity for each potential starting point. Each of these students recognized the relevance of the linear search pattern, but many did not demonstrate sufficient flexibility and employed it erroneously. An additional sixth of the students attempted an on-the-fly computation, but yielded a variety of erroneous solutions. Some searched for the smallest gap between neighbouring barriers, and concluded some irrelevant conclusion from its value, whereas others offered some other manipulations with differences of locations of neighbouring barriers.

Less than a ninth of the students (15 out of 143) recognized the relevant observation described above. They all elegantly described their observation, and demonstrated suitable abstraction, but less than a half of them (7 students) properly mapped their suitable idea into a correct variant of linear search.

As we indicated above, we expected difficulties with this non-trivial task. Its solution actually involves not only specific transfer of a rather "close" utilization of a familiar entity, but also non-specific transfer, of problem solving competence that involves a combination of the aspects of *recognition*, *abstraction*, and *mapping*. The notion of cognitive overload was apparent here in a slightly different way than with the previous tasks. It seemed to us that for the vast majority of the students, the "blurred", or vague picture of the task primarily evolved from the variety of elements to consider during the *task analysis*, prior to the algorithmic patterns' selection and manipulation. Many seemed to feel frustrated from being unable to pinpoint the required problem solving characteristics on which to capitalize.

3   Discussion

The findings of the studies presented in the previous section reveal difficulties with the five aspects of transfer introduced in the Introduction: flexibility, embedment, abstraction, recognition and mapping. These difficulties seem to be related, to some extent, to the notion of cognitive load.

Difficulties with *flexibility* were noticed explicitly with the first population, and implicitly with the other populations. Participants demonstrated incompetence in adapting, and slightly modifying an elementary algorithmic pattern for solving the new task at hand.

Difficulties with *embedment* were observed primarily with the third population. Although these participants studied interleaved decomposition of patterns, in both their CS1 and CS2 studies; many of them were unable to assemble basic algorithmic patterns so that pieces of different patterns are embedded together.

Difficulties with *abstraction* were observed with the third and the fifth populations. Although these participants have seen generalizations of specific patterns in their CS1 and CS2 studies, they did not yield the necessary generalization of the previous-current pattern and the linear search pattern.

Difficulties with *recognition* were observed with the second and the fifth populations. The tasks posed to these participants required recognition of two kinds. In the case of the second population, the posed task required the recognition of the basic previous-pattern; in the more subtle case of the fifth population the posed task required the recognition of an insightful observation on which to capitalize. The former kind involved (difficulties of recognition with) specific transfer, and the latter kind involved non-specific transfer. While the latter may be harder to expect from average students, the former should be expected from such students.

Difficulties with *mapping* were observed with the last two (fourth and fifth) populations. The tasks posed to these participants required mapping of recognized patterns in a way that is suitable for the task at hand. Yet, many demonstrated difficulties of properly mapping the pattern they recognized as relevant for the task.

The aspects of flexibility, embedment, and abstraction are indicated by Marshall as essential features of cognitive scheme acquisition and utilization [4]. The aspects of recognition, abstraction, and mapping are indicated by Gick and Holyoak [2], Sternberg [9], Mayer and Wittrock [5], and others, as essential components of transfer.

A primary cognitive theme related to the above aspects is that of cognitive load. In each of the studies presented in the previous section, it seemed that obstacles faced by the participants were related to "blurred" and vague pictures they had about relevant data in the tasks to be solved, about proper utilization of basic algorithmic patterns, and about interconnections between these patterns. It seemed that knowledge organizations and problem solving resources of many participants were deficient. We believe that CS students and teachers may reach higher levels of competence, with respect to the transfer aspects indicated above, by improved CS teaching practices.

CS instructors may aim beyond ordered presentation and practice of algorithmic, or programming patterns. While explicit instruction of such patterns may indeed reduce cognitive load, the way in which such instruction takes place is essential. It may be made more instructive. Transfer-oriented instruction, which explicitly addresses the five aspects discussed in this study, may further develop and improve student design skills.

Sweller and others suggest careful instructional design [7,10], in which the main premise is to reduce cognitive load on the working memory. This may primarily be achieved by carefully selecting suitable assignments and worked-out examples upon teaching. In addition, one may employ 'guided discovery' [5,7], which can assist integration of new data, and enhance cognitive connections.

We offer an approach of *transfer-oriented instruction*, for enhancing students' design skills. First, instructors should be aware of the five transfer aspects displayed and discussed in this study. Then, we suggest that instructors embed in their teaching explicit activities that develop competence with respect to transfer aspects, as follows.

- Design transfer-oriented tasks and worked-out examples, and embed them throughout the current teaching materials. The tasks and examples will be used both in class and as homework assignments.

- In addressing recognition and mapping, devise and display a set of tasks, and do not always require these tasks' solutions. Rather, ask the students to group tasks together into subsets, according to their solution structures. These structures should encapsulate diverse composition forms of different patterns. You may as well pose tasks such as the one in our fifth study, and elaborate the recognition-of and capitalization-on task characteristics.

- For addressing and enhancing flexibility, devise tasks that require "local" modifications of a just-taught solution scheme, or pattern, as displayed in our first study. In a following class discussion, compare between the task solutions and the just-taught pattern, and explicitly indicate the modifications.

- In addressing and enhancing embedment, devise tasks such as the one in our second, third, and fourth studies, which require interleaved embedment of two or more patterns. Use the terminology of embedment; explicitly underline pieces of embedded patterns and discuss which code pieces belong to which patterns. Students should also be encouraged to offer alternative embedment structures.

- The notion of abstraction is addressed by CS instructors. Yet, one may extend that. Embed assignments in which basic patterns should be used with higher-level entities, as shown in our third and fifth studies. In a following discussion, explicitly display the correlation between a basic pattern and its higher-level application.

- Devise activities that combine a couple of the above transfer aspects (as in our third and fifth studies), and refer to the combined aspects in a corresponding class discussion.

The above guidelines elaborate the focus on structural features of tasks and solutions. Guided discovery and reflective class discussions, in which students compare their thought processes, may be most beneficial [5], and offer instructive illuminations. Pattern invocation, manipulation, and utilization may become more effective, upon learning to sift out relevant from irrelevant information. Future studies (including one that we are currently conducting) may examine the effect of various applications of the above recommendations, in improving transfer during program design.

References

1. Astrachan, O., Berry, G., Cox, L., Mitchener, G.: Design patterns: an essential component of CS Curricula. In: SIGCSE 1998, pp. 153–160 (1998)
2. Gick, M.L., Holyoak, K.J.: Schema induction and analogical transfer. Cognitive Psychology 12, 306–355 (1983)
3. Linn, M.C.: The cognitive consequences of programming instruction in classrooms. Educational Researcher, 14–19 (1985)
4. Marshall, S.P.: Schemas in Problem Solving. Cambridge University Press, Cambridge (1995)
5. Mayer, R., Wittrock, M.: Problem-Solving Transfer. In: Berliner, D., Calfee, R. (eds.) Handbook of Educational Psychology, pp. 47–62. Erlbaum, Mahwah (2006)
6. Muller, O., Haberman, B., Ginat, D.: Pattern-oriented instruction and its influence on problem decomposition and solution construction. In: Proceedings of ITiCSE 2007, pp. 151–155 (2007)
7. Paas, F., Alexander, R., Sewller, J.: Cognitive load theory and instructional design: recent developments. Educational Psychologist 38(1), 1–4 (2003)

8. Spohrer, J.C., Soloway, E., Pope, E.: A Goal/plan analysis of buggy Pascal programs. Human-Computer Interaction 1(2), 163–207 (1985)
9. Sternberg, R.J.: Metaphors of Mind: Conceptions of the nature of Intelligence. Cambridge University Press, Cambridge (1990)
10. Sweller, J.: Cognitive load during problem solving: effects on learning. Cognitive Science 12, 257–285 (1988)

Introductory Computing: The Design Discipline

Viera Krňanová Proulx[*]

Northeastern University, 360 Huntington Ave.,
Boston, MA 02115, USA
vkp@ccs.neu.edu

Abstract. The goal of this paper is to present in context the key didactical principles behind the *Program by Design* curricula, motivate the need for the supporting software, and describe in detail the *How to Design Classes* component for teaching introductory object oriented program design using Java and Java-like languages. The key innovations are a systematic test-first program design, and the introduction of programming language concepts by designing abstractions based on existing programs.

Keywords: Informatics in primary and secondary education, design principles, software for novice programmers, abstractions.

1 Introduction

What is computing? What is informatics? The answer to this question guides the design of the curriculum that focuses on the principles, not fads. At the heart is the computation: a program that consumes data and produces new data according to some formula. But this is just basic algebra, automated. To extend this notion of computation, we need to deal with more complex data. No algorithm exists apart from data. What comes first? We believe that understanding data, how information can be represented as data, and how data conveys information is at the heart of computing and deserves a serious place early in the curriculum. Well-structured data reveals clearly numerous algorithms for extracting new information, and provides the context for learning the foundations of program design. The key questions: the design of abstractions, the concerns about efficiency, the multiple ways the same information can be represented as data, the difficulty of reliable and secure data transmission, the communications protocols, and many others arise naturally in this context.

The traditional curricula for introductory programming start by designing algorithms and overwhelming the student with complex syntax and language features, but providing little guidance on what the program design process should be. Out of more than 20 Java-based textbooks only two mention testing of students programs, and even then without the appropriate software support [1], [13]. Other recent approaches use game-like environments to make the programming more attractive and accessible [3], but still fail to focus on the design process that guides the program design. Tinkering, trial and error approach rules.

---

[*] Partial support for this project has been provided by the two NSF grants DUE-0618543 and DUE-0920182.

I. Kalaš and R.T. Mittermeir (Eds.): ISSEP 2011, LNCS 7013, pp. 177–188, 2011.

In the first part of this paper we describe the key ideas of the *Program by Design* project that introduces systematic program design principles for students ranging from grades 6-7, all the way to the university level. The various components of the *Program by Design* curriculum have been developed over the years by a team of programming language researchers, software developers, and educators. The author has had a key role in the design and implementation of some of the libraries, and in the design and implementation of the most advanced component of this curriculum, formerly known as *ReachJava*.

In the second part we focus on the *ReachJava* component, that presents the design of programs in object-oriented Java-like languages. We describe the role of supporting libraries that expose to the novice programmer the essential design principles while hiding the confusing detail. In the third part we show how this approach leads naturally to a systematic design of abstractions and provides the context for understanding more complex programming language features, as well as the design and the use of standard libraries.

2 How to Design Programs

The foundation of *Program by Design* (known as *TeachScheme!*) has been presented in the textbook *How to Design Programs* [4, 5] and its German language counterpart *Die Macht der Abstraktion* [8]. During the past five years, the curriculum has been adopted and augmented, with dedicated software support, to target young children, ages 10-13. The *Bootstrap* curriculum has been taught to hundreds of children and all materials are available online [1]. The *How to Design Programs* curriculum is appropriate for all students in secondary schools and universities, regardless of their field of interest. While the context of these curricula is the design of programs, the ultimate goal is to teach the students fundamental skills for solving complex problems and organizing the solution in a systematic way.

Functions and Algebra: Bootstrap. Typical first programs students often encounter involve designing and evaluating a simple algebraic function: compute where will a cyclist be after the given time elapsed, if he is traveling at the speed of 25 km/h. We see that the distance is a function of time and can be written as *distance = fnc(time)*. We can explain this idea through simple tables: at times 0, 1, 2, 3, the cyclist will be 0, 25, 50, 75, km away. But with the right programming environment, we can turn these functions into controls of an interactive animation: the movement of the cyclist is represented as a function that for each tick of the clock produces the current location of the cyclist on the screen. Now, what if the response to the *left* and *right* arrow keys that moves another object horizontally is encoded as another function. We add the detection of a collision as a third function, and we have finished programming the model of an interactive game. This is the beginning of the *TeachScheme!* curriculum and is the key feature of the *Bootstrap* curriculum. Children in the Bootstrap program write down the list of locations where the falling ball will be after each tick of the clock, then design the functions that model the movement. The basket catching the ball at the bottom moves in response the keys pressed. The conditional (a function that produces a `boolean` value) is used to update the score. The image of a ball or of a basket is a simple primitive data item in the program. The drawing of the images on the `canvas` (the game board), the invocation of the event handlers (the functions `on-tick` and `on-key` defined by children) and the entire animation is controlled by the provided library.

This is serious work. Children are true designers, learning basic algebra to implement their games. After nine lessons they can explain the evaluation of expressions, the substitution principle, the conditionals, and proudly show their game.

Taking Design Seriously. The didactical principles of the Program by Design curriculum are based on enabling the learner to master a systematic approach to problem solving by following a well-structured design process encoded in three *design recipes*. The *design recipes* give the instructor a tool to diagnose the student's problems by identifying the step in the design process in which the student encounters difficulties.

When we teach children to design functions, we give them a blueprint, a roadmap that shows them the steps in the design process. Once we have identified the data needed to represent both the inputs, and the expected result, we follow the ***design recipe for functions/methods***:

- Write down in English the purpose statement for the function/method, describing what data it will consume, and what values will it produce. Add a contract that specifies the data types for all inputs and the output.
- Make examples of the use of the function/method with the expected outcomes.
- Make an inventory of all data, data parts, and functions/methods available to solve the problem.
- Now design the body of the function/method. If the problem is too complex, use a *wish list* for tasks to be deferred to helper functions.
- Run tests that evaluate your examples. Add more tests if needed.

The children's version is adapted to their abilities, but the focus on systematic design remains. The comment from a child *'I never knew I could divide a big problem into smaller ones'* affirms that these design principles transcend computing and programming. Seasoned programmers recognize that we practice ***test-first design***.

Understanding Data. After the first brief introduction to representing simple programs as functions (that correspond directly to mathematical functions) the *How to Design Programs* curriculum focuses on understanding the complexity of data, the way how information can be represented as data, and, conversely, how data can be interpreted as the information it represents.

The first step in designing a program is always the design of data that represents the problem. The ***design recipe for data definitions*** guides the students as follows:

- Can you represent the information by a primitive data type?
- Are there several related pieces of information that describe one item? If yes, design a composite data type (*struct, class*).
- Does the composite data type contain another complex piece of data? Define that data type separately and refer to it. (A `Book` data item contains an `Author` data item.)

- Are there several variants of the information that are represented differently, but are related (*e.g. a circle, a rectangle, a triangle --- all are shapes*)? If yes, design a union type. (In Java, define a common `interface`.)
- Repeat these steps. This may lead to self-reference, mutual reference, and eventually to a complex collection of classes and interfaces.
- Make examples of data for every data type you design.

Students learn to design complex data: ancestor trees (with person's mother, father, their ancestors); data that represents files and directories in a computer system; ice cream cones with the cone and a list of toppings; a river system with confluences and tributaries; etc. When designing functions for such complex data, the inventory step of the design recipe calls for identifying not only the function inputs, but also the parts of any composite data (*struct, class*), variants of a union type, as well as all functions that are already available for either the input data or the parts of the input. So, if one of the shapes is a combination of the *top* and the *bottom* shape, any function defined for shapes can be used for both the *top* and the *bottom* parts.

Simple language, complex data, serious program design is our motto. All of this can be taught in the context of a very simple language that supports only the appropriate data definitions (with their constructors, selectors, and predicates that identify the data type) and on the functional side provides the standard arithmetic, relational, and logical expressions, and a conditional. If every function produces a new value, the result, then the entire design process is very straightforward:

- Tests are simple, as they only verify that the result matches the expected value.
- Function composition comes naturally, result of any function application can be used in further computations.
- The order of computation does not affect the result. (However, a function or a data item must be defined before it can be used.)

To provide fun and challenge, we provide libraries that handle interactive graphics back ends of game, with students designing the model: the functions that produce a new scene in the game in response to a key event, or timer tick. Drawing scenes using shapes and images is supported through functions that support the composition of images. Games like *pong, snake, space invaders*, provide a design playground.

Designing Abstractions --- Advanced Programming. After we have written several programs (functions/methods) that solve similar problems we begin to see patterns: the solutions are very similar to each other. Students see that certain functions appear similar, the way the data is handled follows the same pattern, or that some code needs to be repeated. To simplify the code and to eliminate repetition, students see the need for more complex programming language features. Rather than using existing libraries to illustrate the generalized solutions, our goal is to teach students how the libraries are built. To achieve this goal, we present a systematic design process encapsulated in the ***design recipe for abstractions*** that helps us eliminate code repetition and produce a more general solution:

- Mark all places where the similar code segments differ.
- Replace them with parameters and rewrite the solution using them as arguments.

- Rewrite the original solutions to your problems by invoking the generalized solution with the appropriate arguments.
- Make sure that the tests for the original solution still pass.

The *How to Design Programs* curriculum now follows with the introduction of local variables, functions as function arguments, mutation of data, as well as the discussion of the efficiency of computation, and additional more advanced topics.

The three *design recipes* are at the heart of the *Program by Design* curriculum. They embody the core questions all programmers face and give the student a guide through the design process. They correspond to the three cornerstones of our curriculum: understanding the connection between information and data and the importance of the design of complex structured data, using the test-first design process for the design of every function or method, and understanding the process of abstraction that turns a problem-specific solution into a generalized solution applicable to a collection of related problems.

3   How to Design Classes

The *Program by Design* curriculum has as its goal to provide a systematic introduction to the fundamentals of computing and programming. The ideas introduced at the beginning apply equally well in a more complex context. The *ReachJava* component with the draft of a textbook *How to Design Classes* [4] extends the original *TeachScheme!* curriculum to the context of class based programming using Java-like languages by introducing most of the essential concepts of object-oriented program design. It is appropriate for secondary schools and universities.

The goal of this section is to reflect on what we learned during the last nine years of implementing the *ReachJava* curriculum and designing the supporting software. We start by showing how the pedagogical principles of *Program by Design* imply the need for novice-appropriate software libraries that support this methodology. We then show how the *ReachJava* curriculum teaches students through systematic design of abstractions to build reusable software and to use standard software libraries.

Libraries for Novice Programmers: FunJava. While many functional languages (such as Scheme) have a compact and fairly simple syntax (at least at the beginner's level), statically typed object-oriented languages such as Java or C# require a complex syntax for solving even the simplest problems. To eliminate a number of problems novices face, our curriculum starts with a limited version of Java (*FunJava*). There is no assignment statement, all fields are initialized either when defined, or in the only constructor allowed. A `class` can implement only one `interface`, and there are only two statements: `if` with a required `else` clause, and `return expression`. This enforces a mutation-free programming style, the original goal of the designers of object-oriented languages. Every method produces a new value, a new instance of data. Rather than starting with *algorithms*, we first practice designing classes and collections of classes and interfaces that represent different, gradually more complex, information. Students design classes that contain fields that are instances of another class, unions of classes, self-referential data, mutually-referential data. The earlier examples of data: ancestor trees, a model of a river system with a number of confluences, the representation of computer files and directories (that contain other files and directories), the representation of a route through the cities, a student's record with the list of courses she is enrolled in, now define a collection of interconnected classes and interfaces.

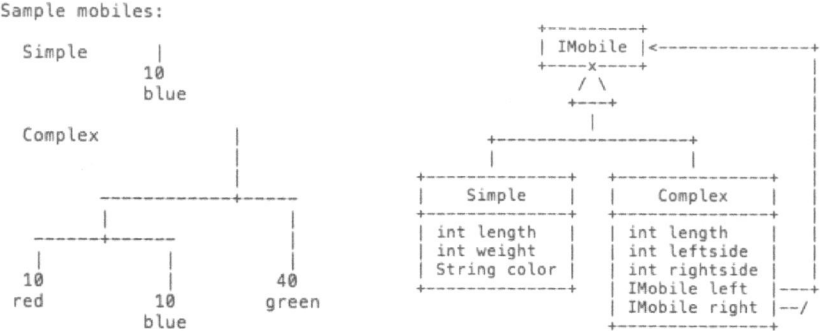

Fig. 1. The examples and a class diagram for a program that models a hanging mobile

Libraries for Novice Programmers: Tester Library. The *design recipe for data definitions* guides the design decisions and teaches a systematic approach to understanding the complexity of data. We use a simple version of class diagrams to illustrate the relationships between classes and interfaces: the containment and the inheritance.

Once we have examples of classes and data, we turn to designing methods, following the same *design recipe*. Functions become methods, and the object that invokes the method (this) becomes just an additional argument the method consumes. Without mutation, the outcome of every method depends only on its inputs. So the students only need to check that the outcome of a method invocation produces the desired value.

```
// is this mobile balanced?
boolean isBalanced(){
        return this.left.totalWeight() * this.leftSide ==
               this.right.totalWeight() * this.rightSide &&
               this.left.isBalanced() &&
               this.right.isBalanced();
}
```

Fig. 2. Sample method in class Complex for a program that models a hanging mobile

Here we encountered a problem: Java and most object-oriented languages do not support equality comparison based on the value of data, and so the design and evaluation of tests in this context becomes a daunting task. We have solved this problem by designing the *tester* library [7], [8], [9] that compares any two objects by the value of their fields, traversing deeply to the primitive components, detecting circularity of data definitions, and making the test design simple and straightforward.

When the tests are evaluated, the student may choose to pretty-print all data fields defined in the `Examples` class, the class that represents the client to the student code, and to print either all test results, or only those that have failed.

```
IMobile big =
     new Complex(1, 6, 6,
                 new Simple(1, 15, "blue"),
                 new Complex(2, 6, 4,
                             new Complex(3, 2, 4,
                                         new Simple(1, 10, "green"),
                                         new Simple(3, 5, "red")),
                             new Simple(2, 20, "red"))));
// Tests for isBalanced
boolean testIsBalanced(Tester t){
    return (t.checkExpect(simp1.isBalanced(), true) &&
            t.checkExpect(comp1.isBalanced(), true) &&
            t.checkExpect(big.isBalanced(), true) &&
            t.checkExpect(comp2.isBalanced(), false));
}
```

Fig. 3. The examples of data and tests for a program that models a hanging mobile

Libraries for Novice Programmers: World Game Library. One may ask, what kind of programs can students write in such a simple environment? Well we can design binary search trees, programs that represent cells in a spreadsheet that refer to other cells with formulas that need to be evaluated, build recursively defined lists of items, thus implementing a stack data type, tennis tournaments, etc. But to support design explorations and to motivate students, we have also built a *world* library [7] that allows students to program the behavior (the model and the display) of a graphics-based interactive game. Students extend the `World` class by adding fields that represents various game objects. They define the methods that represent the actions in response to the timer or a key press, producing a new instance of a changed world, and the methods that produce the scene that represents the current state of the world. The library creates the game canvas in a new frame, installs the necessary event listeners, and provides event handlers that invoke student-defined methods.

We can accomplish a lot with simple tools. The three libraries: *FunJava* that provides a novice-friendly simple language, the *tester* library, that makes the test design, method evaluation, and data display easy, and the *world* library that turns simple programs into interactive graphics-based games provide the infrastructure where student's focus is on the program design, free of idiosyncrasies and complexities of professional programming languages and libraries.

The great advantage of this approach is that the students learn to program in a truly object-oriented style from the beginning. They understand the dynamic dispatch of methods. We insist that every method handles only one task and delegates to helper method any complex tasks that arise (*the chain of responsibility principle*). Students have to reason about which class needs to be responsible for every task (i.e., where should the methods be defined), and they have to write examples of method invocation with the expected outcomes (tests) for every method they define.

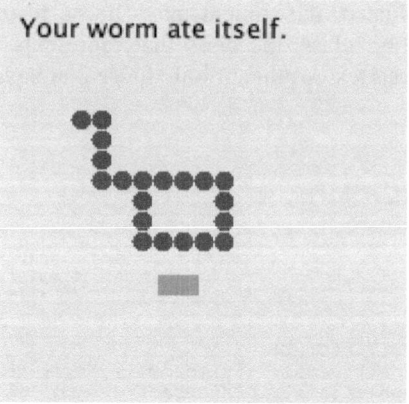

Fig. 4. The snake game (by Matthias Felleisen). The snake moves on each tick, changes the direction in response to the arrow keys, looks for food, grows with each food eaten, avoiding the walls or itself.

With this foundation, we are ready to discuss more advanced ideas of program design and introduce programming language features that enable the design of reusable libraries. Each new programming language feature is introduced in the context of solving a problem encountered earlier: the way to eliminate code repetition, the way to handle problems that cannot be solved using purely functional style, the way to eliminate the need for excessive saving of intermediate results, etc. The framework for this stage of the curriculum is the study of designing *abstractions*.

4   How to Design Libraries

A novice has a hard time learning a number of features of modern object-oriented languages that have been designed to help a seasoned programmer to work effectively. It is important to present every language feature in the context of compelling examples that illustrate the reason for introducing that feature. Once our students mastered the basics of the program design in the object-oriented style, we focus on the design of abstractions that leverage different language features to avoid code repetition and to build reusable code. This provides a context for learning how libraries are designed and used. The ***design recipe for abstractions*** provides a systematic way to examine where abstraction is possible, to define what needs to be done, and to verify that the abstraction correctly accomplishes the desired task.

Abstract Classes. In the introductory weeks students have seen several classes (Circle, Square, Rectangle) that implement the common interface Shape. Each class included a field that represented the location of the shape in some *Canvas*, and it included methods that compute the area of the Shape, it distance to the origin, and a method isSmallerThan that compared the area of **this** Shape to the area of the given Shape. The code repetition in these classes is obvious, and it is easy to motivate the need for an abstract class that defines all common fields, includes a constructor that initializes them, and contains a concrete implementation of the common methods as well as those that are common to most subclasses.

The other side of this coin is the introduction of our abstract World class that provides the entire functionality for designing an interactive graphics-based game, leaving to students the task of implementing the abstract onDraw method, and overriding the stubs of the onKey and onTick methods. This is their first encounter with a library. It provides an environment for designing interesting applications while focusing only on the design of the model.

Function Objects. *Java Collections Framework* includes several interfaces that specify functional behavior. The most commonly used ones are the Comparable interface and the Comparator. The only role the Comparator plays is to provide a wrapper for a method compare that compares two objects of the same type. In many functional languages, a function can be passed as an argument to another function (*functions are first class values*), but Java designers did not provide for this. Thus the programmer needs to define a class that implements this interface, design the needed method, define an instance of this class and pass that as the argument to the methods like sort or findMin, or findMax. However, rather than introducing these interfaces, we start with interfaces ISelectBook or ISelectPerson that implements a predicate that selects the objects with the desired properties and is used by methods like findBook, containsPerson, etc. The reason is to delay introducing the type parameters, and to show the students both the definition of the interface and the design of the classes that implement it. One example we use is to select all runners in the Boston Marathon that are female under 40 years old, all masters runners (over 50) etc. It is clear that we do not want to design the same selectRunners method several times, when the only thing that changes is the selection criterion.

We do mention that a similar technique is used for defining the action that the computer performs in response to the GUI button press or when an event handler is activated by the event it is listening to.

Mutation (State Change). We introduce the assignment statement and the resulting change of the state of a variable once the students are comfortable with designing classes, designing methods, and they understand the dynamic dispatch of the methods. We present problems that either cannot be solved without mutation (or some additional language construct), or where mutation simplifies the work to be done. One cannot design the data that represents students enrolled in courses, when the course data contains the list of currently enrolled students, and the student data contains the list of courses student is enrolled in, without changing the values of the data fields after they have been defined. A bank record representing an account needs to be modified when a deposit or a withdrawal is made, so that every program that has access to this data sees the change. The variable that holds the user's response to a question will only get its value once the user responds.

A survey of typical textbooks and papers describing introductory curricula shows that only a handful of them pay any attention to systematic design of tests from the beginning. We attribute this to two problems: the design of tests in the presence of state change is quite complex, and the design of test for mutation-free programs requires support for extensional equality tests. Yet, designing programs that are not tested is a very bad habit to learn. Our experience has repeatedly shown us that even seasoned programmers make trivial mistakes in the simplest program components and that these are either very hard to find, or go undetected for extended periods of time.

Our introduction to state change comes hand-in-hand with the design of tests: with the setup of needed data, method invocation, testing of the effects of the method, and the reset of the data that has been used. The purpose statement for the method changes to include the word *EFFECT* where necessary. We defer until later the use of methods that combine the state change with returning a new value.

This is also the time when we begin to discuss the difference between two objects that represent the same value and two names for the same object. By now students are comfortable with the basic program design, the language syntax, and they can appreciate the subtleties of the data representation and aliasing.

Program Integrity and Usability. At this point students are ready to think of programs that will be used and modified by others. We introduce several techniques a programmer can use to expose the program behavior while hiding and protecting the internal details of implementation. We talk about the *visibility modifiers*, show how *constructors* can provide several different ways for the user to instantiate objects and to verify that the data satisfies the desired constraints (month is one of 12 possible values, hour does not go beyond 24, etc.), and introduce the *exception handling*.

Another important topic we begin to discuss is the definition of *equality* and the implementation of methods that compare two objects. Are two lists the same if they refer to the same instance? Or are they the same if their respective elements refer to the same instances or just represent the same values? The need for detecting circularity in data also comes into play.

Parametrized Types. Students see that we have been defining similar methods for data collections of different data types: binary search trees of persons, cities; lists of books or songs, etc. Even the function objects were targeted for only one type of objects. Having seen this, the introduction of generics (parametrized types) is a welcome new abstraction in spite of the complexities of the necessary syntax.

Abstracting over Traversals. All along we also present examples of methods that represent traversal over the items in the collection of data. We design methods isEmpty(), Data getFirst(), and Collection getRest() for both lists and binary search trees. Abstraction over the collection of these three methods introduces a new interface, Traversal, that represents a functional iterator:

```
Interface Traversal<T>{
  public boolean isEmpty();
  T getFirst();
  public Traversal<T> getRest();
}
```

We see that the methods that manipulate the collections of data can be defined outside of the class definitions of these collections. We have come a full circle: starting from standalone functions in the functional language, to designing methods that rely on the dynamic dispatch for selecting the appropriate action, to moving the methods to the

Algorithms class that deals with an arbitrary data, as long as the data collection provides the necessary hooks.

Loops, the *Java Collections Framework* Iterator interface, and the Iterable interface are introduced at this time.

Abstract Data Types. With this background we introduce the ArrayList, the HashMap, the Stacks and Queue, and other classes in the *Java Collections Framework*. We ask the students to implement the Stack and the Queue interface; design a mutable linked list, and use them in the context where one or the other can be used interchangeably. The *Depth-First Search* and the *Breadth-First Search* over graphs differ only in the way we implement the data set that keeps track of the next set of edges to consider.

When introducing the hash maps, we revisit the issue of equality. We show how to correctly override both the equals method and the hashCode method. Using the *JUnit* test framework, reading and writing the *Javadoc* style documentation are the last couple of steps for students to be ready to fully use the standard Java libraries.

Java Collections Framework. With students' knowledge of the meaning of interfaces for defining the behavior of data, abstract classes for implementing the common behavior of a union of similar data types, the use of function objects to define functions that algorithms can use, the introduction of the *Java Collection Framework* is very straightforward. Students understand the design, can reason about the implementation, and can implement some of the library classes themselves.

We complete the work with several discussions of the **resource management issues**. Memory usage, time-complexity of algorithms, the cost of using structural recursion, all are made visible through an assignment where students evaluate stress test runs. The classical data structures and algorithms are presented only to illustrate the design choices: indexed data structures make binary search possible, key-value associations allow for fast data lookup, linked lists allow localized modification of the structure, quicksort leverages the divide and conquer strategy, etc. Through simple programming assignments we show students the different ways how information can be represented as data: students manipulate images by modifying image pixels, they process text data computing word frequencies, they use our simple sound library to generate sound effects and background music for their games.

5 Summary

The *Program by Design* curriculum evolves. The *Bootstrap* component is building a web-based programming environment [10], the second edition of HtDP includes support for client-server computing over the network [3]. We have piloted a library that supports the design of applications for mobile devices. The tester library is a foundation for the development of a comprehensive software testing curriculum.

The curriculum has been used in many settings (after-school programs, summer camps for children, secondary schools, universities). Teachers of children who completed the *Bootstrap* program wonder at their improved math grades. Secondary school students who started with the *Program by Design* curriculum do well in the Advanced Placement in Computer Science (AP) test, even though the AP curriculum follows a more traditional programming curriculum. Our university added a required course for the graduate Master's of Science program that is based on the *Program by Design* curriculum, to improve advanced student's program design skills.

Acknowledgments. The *Program by Design* is a work of the team led by Matthias Felleisen, with Matthew Flatt, Robby Findler, and Shriram Krishnamurthi its co-founders [7]. Kathy Gray has contributed to the design and initial implementation of the ReachJava segment [6, 11]. Kathi Fisler has worked on the further development of the curriculum. Emmanuel Schanzer is the designer of the *Bootstrap* component [2], [14]. Erich Neuwirth inspired the development of the sound library [9].

The two grants by the National Science Foundation (Redesigning Introductory Computing: The Design Discipline, DUE-0618543 and Integrating Test Design into Computing Curriculum from the Beginning DUE CCLI 0920182) provided partial support for the development and dissemination of this project.

References

1. Barnes, D.J., Kölling, M.: Objects First with Java: A Practical Introduction using BlueJ. Prentice Hall / Pearson Education (2008)
2. Bootstrap Project, http://www.bootstrapworld.org
3. Dann, W.P., Cooper, S., Pausch, R.: Learning to Program with Alice, 3rd edn. Prentice Hall, Englewood Cliffs (2012)
4. Felleisen, M., Findler, R.B., Flatt, M., Krishnamurthi, S.: How to Design Programs. MIT Press, Cambridge (2001)
5. Felleisen, M., Findler, R.B., Flatt, M., Krishnamurthi, S.: How to Design Programs, 2nd edn., http://www.ccs.neu.edu/home/matthias/HtDP2e/index.html
6. Felleisen, M., Findler, R.B., Flatt, M., Gray, K., Krishnamurthi, S., Proulx, V.K.: How to Design Classes, http://www.ccs.neu.edu/home/matthias/htdc.html
7. Findler, R.B., Flanagan, C., Flatt, M., Krishnamurthi, S., Felleisen, M.: DrScheme: A pedagogic programming environment for Scheme. In: Hartel, P.H., Kuchen, H. (eds.) PLILP 1997. LNCS, vol. 1292, pp. 36–388. Springer, Heidelberg (1997)
8. Klaeren, H., Sperber, M.: Die Macht der Abstraktion, B. G. Teubner Verlag, Wiesbaden (2007)
9. Neuwirth, E.: http://sunsite.univie.ac.at/musicfun/MidiCSD/
10. Proulx, V.K.: ReachJava Libraries, http://www.ccs.neu.edu/javalib
11. Proulx, V.K.: Test-Driven Design for Introductory OO Programming. SIGCSE Bulletin 41(1), 138–142 (2009)
12. Proulx, V.K., Gray, K.E.: Design of Class Hierarchies: An Introduction to OO Program Design. SIGCSE Bulletin 38(1), 288–292 (2006)
13. Riley, D.D.: The Object of Data Abstraction and Structures Using Java. Addison Wesley, Reading (2003)
14. WeScheme, http://www.wescheme.org/

A Short Introduction to Classical Cryptology as a Way to Motivate High School Students for Informatics

Lucia Keller, Barbara Scheuner, Giovanni Serafini, and Björn Steffen

Department of Computer Science, ETH Zurich, Switzerland
{lucia.keller,barbara.scheuner,giovanni.serafini,
bjoern.steffen}@inf.ethz.ch

Abstract. In Swiss high schools, programming is the typical content of an introductory informatics course. This is an important topic, but nevertheless it is only a part of the field. By integrating short introductions to other topics, students get a better understanding of the broadness of informatics.

This article presents such a short introduction unit about classical cryptology without requiring any school-related prior knowledge in informatics. The basis of this unit is the everlasting game between code designers and code breakers to build, respectively break, cryptosystems. The challenge of breaking the codes presented by the teacher is the core and motivating factor of our didactical concept. Although the theoretical concepts cannot be presented in detail, the unit demands analytical skills and encourages critical thinking.

The unit motivated 70 % of the participating students to learn more about the topic, which is a good pre-condition for subsequent cryptology courses.

1 Introduction

Cryptology is nowadays an interdisciplinary research field, which integrates elements of mathematics [1,3], algorithmic fundamentals of computer science [8] as well as physics [4]. Moreover, cryptology is an exciting school subject, which allows the combination of the student's real-life experiences with deep scientific knowledge. Students of different ages and with different abstraction skills can be introduced to the basic mechanisms of cryptology.

Koblitz stated already in 1997 the impressive value of cryptography as a teaching tool. In his article [12] he highlighted his didactic approach and his experiences teaching symmetric and public-key cryptography concepts to children of primary schools. Bell et. al. devoted some of their off-line activities and games to cryptology in *Computer Science Unplugged* [2]. Also the very popular book of Singh [13] shows that the history behind cryptology can attract a large audience.

Our paper relies on a similar didactic concept, but describes a teaching sequence especially for high school students, which only focuses on classical cryptography and, therefore, handles more cryptosystems than Koblitz did. This

I. Kalaš and R.T. Mittermeir (Eds.): ISSEP 2011, LNCS 7013, pp. 189–200, 2011.

teaching unit does not require any school-related prior knowledge in informatics. It is designed as an interaction between the code designer (the teacher) and the code breakers (the students). While the mere encryption of a given plaintext and the decryption of an available cryptotext soon becomes annoying, students challenged to break cryptosystems are really involved and implicitly adopt a critical attitude regarding scientific subjects.

The introduced teaching unit on classical cryptology is a part of a comprehensive collection of teaching materials for high schools developed at ETH Zurich during the last five years. The authors attach importance to a precise and understandable explanation of concepts and notions. This is an important basis for a profound comprehension of the design and the analyis of the introduced cryptosystems. Those books are self-contained and can be used for individual learning [5]. The basis of the presented teaching unit are chapters 1 to 3 (90 pages) of the textbook about cryptology. The whole book is covering topics such as classical and modern cryptology, symmetric and public-key cryptosystems, zero-knowledge protocols as well as their mathematical and algorithmic fundamentals [6]. This teaching unit was intensively tested in some informatics classes and during several project weeks or single visits aiming to promote informatics at high schools. For a thorough discussion of those chapters we allocate 18 lessons in a informatics class in the last year of high school.

The goal of our teaching unit is not a thorough discussion of this topic but to give the students some interesting insights to classical cryptology without using the formal language of mathematics too intensively. We want to elate the students for informatics and to motivate them for visiting further informatics courses. Therefore, we use classical cryptosystems, although they are far from satisfying modern security requirements. But they represent an essential didactic tool in order to introduce the basic concepts and the well-established, precise terminology of cryptology.

The paper is organized as follows: in Section 2, we introduce the concepts of classical cryptology as well as the basic terminology. After that, we describe the main idea, the goals and the structure of the teaching unit. Section 3 presents a sample lesson held at the beginning of 2011 at the high school MNG in Zurich and discusses the results of a survey the students had to complete in conjunction with the sample lesson. We conclude with general reflections on our experiences dealing with motivation, expectations and achievements introducing high school students to classical cryptology in Section 4.

2   Classical Cryptosystems and the Concept of Security

It is important to determine the terminology before starting to introduce examples of cryptosystems.

We consider *classical cryptosystems* as symmetric (secret-key), paper-and-pencil cryptographic systems, which were conceived and employed in the pre-computer era.

A person, called the *sender*, wants to send a message (the *plaintext*) in a natural language to a second person, the *receiver*. The plaintext will be sent over an insecure channel. Hence, the message may be eavesdropped by an unauthorized person. Therefore, we have to *encrypt* the message with a secret key such that we can send the *cryptotext* over the insecure channel. The receiver *decrypts* the message with a method that inverts the encryption. Such cryptosystems are called *symmetric cryptosystems* because the sender and the receiver encrypt and decrypt with the same secret key.

2.1 Security of Cryptosystems

It is naive to assume that it is adequate to keep the method of *encryption* and *decryption* secret between the sender and the receiver. In 1883, Kerckhoffs formulated the so called *Kerckhoffs' Principle of Security*:

> *A cryptosystem is secure, if one, knowing the art of the functioning of the cryptosystem but not knowing the key used, is not able to derive the original plaintext from the given cryptotext.* ([8], cf. [10,11])

Nowadays, being not able to derive the original plaintext means that the unauthorized person is not able to find the plaintext with his actual computational resources in reasonable time. In some applications, reasonable time means 10 seconds and in other applications it means 30 years. This depends on the importance of the plaintext and the time how long a cryptotext has to remain confidential. We will see that it is not sufficient for building secure cryptosystems to have a huge number of keys, but there also must not exist an efficient algorithm that can break a cryptotext without the knowledge of the key.

According to Kerckhoffs' Principle of Security, a secure cryptosystem has not to be kept a secret. Also if an enemy knows everything about the cryptosystem, he is not able to decrypt a cryptotext without knowing the key. Hence, the security only depends on the secrecy of the key and the strength of the cryptosystem.

In this paper, the plaintext is encrypted without punctuation marks and spaces. For example, the sentence

> This is a sample sentence.

is transformed to

> THISISASAMPLESENTENCE

before encrypting it.

3 The Didactical Concept of the Lecture

In the following, we describe how we introduce the students to classical cryptology. The presented lesson takes about 90 minutes and is only meant to give an overview of the topic. A thorough introduction would need much more time (approximately 18 lessons) as classical cryptology is strongly related to non-trivial concepts of probability theory and algorithmics which are used for the design and the analysis of the cryptosystems. A detailed description of a longer introduction of classical cryptology can be found in the textbook [6]. A shorter introduction is given by Hromkovič [8,7]. Nevertheless, we focus on clear explanations of concepts and notions as a basis for following lessons.

3.1 Teaching Goals

The principal aim of our teaching unit consists in influencing the attitude and the interest of high school students with respect to cryptology and its applications in real-life: we expect that after completing the teaching unit, the students are aware of the relevance of cryptology for today's life and in particular for secure communication systems. At the same time, we try to stimulate the student's interest on cryptology and informatics in general and possibly to motivate them to learn more about this topic.

Students attending the teaching unit presented in this paper acquire concrete, observable skills and capabilities: they know and understand the concepts of cryptosystems, sender, receiver, eavesdropper, encryption, decryption, secret-key, plaintext, cryptotext as well as the definition of the concept of security for cryptosystems. Students correctly explain all these concepts to school colleagues without prior knowledge in cryptology, in a non-formal way, relying on the introduced terminology and the presented graphical schemes.

Students are able to correctly deal with the presented classical cryptosystems and to encrypt and decrypt simple messages by hand (without a computer). Given a message encrypted with one of the presented cryptosystems, students are able to compute the size of the key-space and to break the encryption carrying out a cryptotext-only-attack, by hand, working alone or in small groups.

3.2 Introducing the Concepts of Cryptology

In the following, we summarize our teaching unit divided in 9 steps.

Step 1: Introduction. We start with a short introduction where we briefly mention some applications of cryptology to motivate the students for these two lessons. Some of them are listed in [8].

Step 2: Two Ciphers. One of the first known ciphers[1] is POLYBIOS. The greek writer Polybios arranged the Greek alphabet, consisting of 24 characters, in a 5x5-matrix from left to right and top to bottom. To encrypt, he replaced every character by a pair, the number of the row and the number of the column.

Another well-known possibility to encode characters is the cipher FREEMASON. It was invented in the 18th century. Every character of the alphabet is replaced by the lines and dots in the neighbourhood of the character (see Fig. 1).

---

[1] Note, that we distinguish the words *cipher* and *cryptosystem*. Cryptosystems use keys for encryption and ciphers do not.

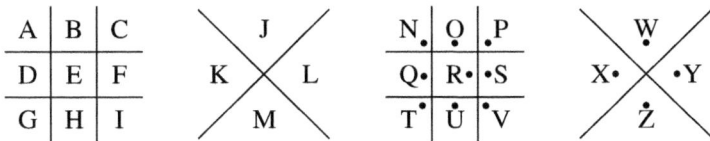

Fig. 1. The FREEMASON cipher [13]. Some examples: A = ⌐, J = ∨, S = ⊏, Z = ∧, and so on.

Step 3: Cryptosystems. At this point, we ask the students why the mentioned ciphers are not practicable and secure. The students immediately see that it is sufficient to know which cipher is used for the cryptotext and then it is easy to decrypt since you can consult the table. Therefore, we need more involved ciphers. Those ciphers should have the property that you do not have to keep the method of encryption secret but only the key. Hence, you need ciphers with a larger variety to encrypt. Those ciphers are called cryptosystems. The two ciphers from Step 2 are not cryptoystems because they do not have keys.

Step 4: The Cryptosystem CAESAR. The probably most popular cryptosystem is CAESAR. In this cryptosystem every character is shifted cyclically by a fixed amount of positions indicated by the key. In the following picture, the key is 2:

Z A B C D E F G H I J K L M N O P Q R S T U V W X Y Z

A B C D E F G H I J K L M N O P Q R S T U V W X Y Z A

This cryptosystem has only 26 keys and hence, is easy to break for an unauthorized person. One can, for example, try exhaustively all keys and if a reasonable text shows up one can assume that the plaintext was found. However, it highly depends on the plaintext and the cryptosystem, if such a brute-force cryptanalysis can succeed in general or not.

Cryptology was and still is a game between the code designers and the code breakers. The **code designers** always try to invent new cryptosystems that are secure and the **code breakers** want to break those systems. We use this game to get the attention of the students. During the whole lecture, the teacher plays the role of the code designer and the students break the systems. This motivates the students because, finally, they can for once correct the teacher. Moreover, the kids also have fun solving puzzles.

Step 5: The Cryptosystem SKYTALE. To get a better cryptosystem, we have to design one with more keys. One possibility is the cryptosystem SKYTALE. The sender and the receiver have a wooden stick with the same diameter. The sender wraps a paper or a leather strip around the stick and writes the

L. Keller et al.

message row by row onto the paper such that he places one character on one winding of the strip. The amount of characters on one winding (the diameter) is the key of the cryptosystem. If the strip is unwrapped, we get the cryptotext. The receiver has to again wrap the strip around the stick in order to read the plaintext. But how many keys do we have? It is easy to see, that this number depends on the length of the text, say n. In this case we have at most n possible keys. But not all keys make sense. In practice, the number of keys is not very large.

Afterwards, we shortly discuss one problem with the cryptosystem SKYTALE. The blanks at the end of the written message give the cryptanalyst a hint about the size of the key. How can we hide these blanks? Usually, the students have some ideas like filling in arbitrary characters at the end of the text, or adjusting the size of the key and the number of windings to the length of the text, such that there will not be any blanks at the end of the text.

Note, that this cryptosystem is based on a different principle than the previous one – instead of replacing symbols we change their positions.

Step 6: The Cryptosystem RICHELIEU. The last simple symmetric cryptosystem we want to introduce is RICHELIEU. This system was invented in the 17th century by the cardinal Richelieu. He used a cardboard where some holes are punched in. Then he writes the plaintext onto the paper through these holes. Afterwards he removes the cardboard and filles in the gaps with arbitrary characters to get the cryptotext. The receiver can decrypt the ciphertext by placing the same cardboard with the same holes on top of the cryptotext. How many keys do we have? Suppose that the number of rows on this cardboard is m and the number of columns is n. Each field on this cardboard can either be punched out or not. Hence, we have two possibilities for every field and $2^{m \cdot n}$ is therefore the total number of keys.

If the students are familiar with combinatorics, then they have no problems to figure out the number of keys. Otherwise, the teacher has to help the students, but also in classes with younger students, it was never a problem to explain how the number of keys is calculated.

Step 7: An Exercise. The Spartans want to send a secret message from Sparta to Athens. In this message, a strategy for the upcoming battle is described. The battle begins in 3 days.

The goal of the Spartans is to encrypt this message in such a way that the enemies need more than 3 days to break it. The Spartans know that the enemies have only one cryptanalyst. He is very smart and works efficiently. For one trial to decrypt the cryptotext, he needs only 1.5 minutes. This trial corresponds to testing one key. The unresting cryptanalyst can work three days and three nights at a stretch. The Spartans have to reckon that the smart cryptanalyst knows which cryptosystem is used for the encryption.

The Spartans decide to use the following cryptosystem:

Cryptosystem 3DAYS
plaintext alphabet: Latin characters
cryptotext alphabet: Latin characters
set of keys: (i, k, j) with $i, j \in \{0, 1, \ldots, 25\}$ and $k \in \{1, 2, \ldots, 100\}$
encryption: Encryption in 3 steps:
                               1. Encrypt the plaintext with CAESAR and key i to $text_1$.
                               2. Encrypt $text_1$ with SKYTALE and key k to $text_2$.
                               3. Encrypt $text_2$ with CAESAR and key j to $text_3$.

Describe the decryption of a message encrypted with 3DAYS. Did the Spartans make a good choice?

Most of the students realize that SKYTALE and CAESAR are independent and that therefore the order of the encryption steps is irrelevant. Further they come to the conclusion that the successive application of CAESAR with keys i and j results in a single CAESAR encryption with the key $i + j \bmod 26$. Hence, the number of different keys is small enough to check all of them within 3 days.

Step 8: A Monoalphabetic Cryptosystem. Now, we look at a better cryptosystem, called the MONOALPHABETIC cryptosystem, it is similar to CAESAR. Again, every character is replaced by a different character, but instead of shifting the characters by a fixed number of positions, we jumble the characters arbitrarily (i.e. we take a permutation of the characters):

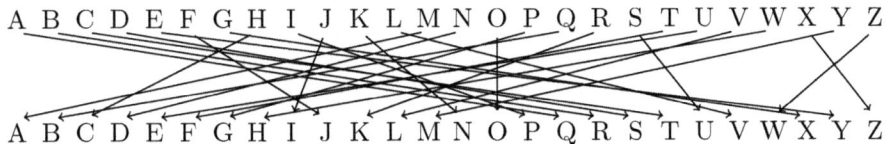

How many keys do we have now? $26! \approx 4.03 \cdot 10^{26}$ and this is a huge number. Nobody is able to check so many keys within reasonable time. But can one exclude another possibility to break a cryptotext that is encrypted with this cryptosystem? Many students discover by their own how the cryptanalysis works: In every language, some of the characters appear more often than other characters. In German, for example, E is the most frequent character followed by N and I. Comparing the frequencies of the letters of a given cryptotext to the expected statistical distribution provides us with a hint for the cryptanalysis. Additionally, we can pay attention to some words that occur often in texts, such as *und* or *das* in German.

It is important though that this frequency analysis only works reliably for large cryptotexts. For the students, a text should be provided that is not too long, but in which the frequencies of the characters more or less coincide with the expected statistical distribution. Tables of these distributions are given for example by Wätjen [14] for German and English.

Step 9: A Longer Exercise. Now, it is time for a longer exercise. The students form groups of two or three people and every group gets a print-out of a table with the expected letter frequencies and a cryptotext, which is a monoalphabetic encryption of a German text. We use the following text [6] at schools[2]:

```
HOQ HQKSD , DOCOPTSPXXO KERJEHOLZOS , PXN PT VOXOS
HOX TOSXLCOS NPOR OPSDOVEQJOIN ; XLCMS HPO
OPSRKLCXNO SOEDPOQ WOQECN UK KER HOQ KEXXPLCN , OPS
VPXXOS JE NOPIOS , HKX KSHOQO ESX YMQOSNCKINOS .
OPSPDO XPSH DIEOLZIPLC DOSED , OPSOS WOQER JE
RPSHOS , HOQ PS HOQ IMOXESD YMS QKONXOIS WOXNOCN .
KWOQ HPO TOPXNOS YMS ESX TEOXXOS HPOXOS HQKSD TPN
HOQ IMOXESD ZEOSXNIPLC JE ESXOQOQ ESNOQCKINESD
KEXDOHKLCNOQ QKONXOIKERDKWOS XNPIIOS .
HONOZNPYDOXLCPLCNOS ESH ZQOEJVMQNQKONXOI VOQHOS
YPOIOS SENJOS ; OPSPDO VOSPDO TMODOS XPLC HOQ
OSNXLCIEOXXOIESD YMS DOCOPTXLCQPRNOS CPSDOWOS .
```

Most of the students need between 20 and 30 minutes to break this text. Those groups are successful, which divide the work in reasonable parts. It is not necessary to count all the characters. Sometimes it is sufficient to determine what E, N and I are and then you may see some patterns from words that you know. The best way to break it is possibly a mixture between counting the characters and searching for known patterns. The students choose this way intuitively.

In this exercise, the students can live out their curiosity for puzzles. We observed that also students that are usually not easy to motivate for informatics like this type of exercise. Because they are in the position of the cryptanalyst and some minutes ago the task seemed to be unsolvable, their ambition to break the cryptotext is really high.

It is important that we do not break off the exercise sessions until at least half of the students finish their work and almost everybody understands how to break the cryptosystems or to solve the exercise. Usually, most of the students finish the cryptoanalysis approximatly at the same time. If this is not the case, you can give the faster students more challenging exercises to solve until the others are finished. Also, if some of the students have already found the solution, the other students are still eager to break the cryptotext on their own. Sometimes we offer prizes for the first groups who break the text but even if there is nothing to win, the students are willing to solve the exercise.

Steps 1 to 6 are strongly teacher leaded parts. The teacher presents the cryptosystems and the students try to find together the weeknesses of the systems. From Step 7 on, the students work for the most time for themselves and in smaller groups. They need much more time for the last three steps.

In most classes, we need 90 minutes up to this point. It is not unusual that some classes are faster. We were able to give them also a short overview of polyalphabetic cryptosystems.

---

[2] To make the task easier, we do not take out the punctuation marks and the blanks.

3.3   Continuing Lessons

The search for a secure cryptosystem continues. The first measure of the code designers was to build cryptosystems with more and more keys. The monoalphabetic cryptosystem has a lot of keys, but we were still able to break the cryptotexts with a frequency analysis. The problem was that the frequencies of the characters in the plaintext are a permutation of the frequencies of the characters in the cryptotext. Is there a possibility to mask those frequencies in the cryptotext? One possibility is to use *polyalphabetic cryptosystems*. One example is the cryptosystem VIGENÈRE. The sender and the receiver must agree on a keyword, which is written from left to right multiple times without a gap underneath the plaintext. Every character of the plaintext is encrypted with the cryptosystem CAESAR: The rank of the character in the alphabet of the keyword underneath the character of the plaintext determines the shift. Hence, it depends on the position of the character in the plaintext, with which shift it is encrypted.

VIGENÈRE was invented around 1550 and named after Blaise de Vigenère. At that time, it was considered as unbreakable. Not until 300 years later, Charles Babbage managed to break this cryptosystem: For a key length n, he divided the cryptotext in n parts such that the characters in every part were encrypted with the same shift. Then, every part can be tackled with the frequency analysis. For this analysis, we have to assume that we know the length of the key. However, there are also methods to determine the key length (see [6,13]).

If the teacher only wants to give an overview to cryptology, he can outline public-key cryptology instead. A teaching unit as well as its scientific background is, for example given by Keller et. al. [9] (cf. [2,12]).

4   Experiences

We were teaching this lesson in about 20 schools in Switzerland. In the majority of cases, the schools invited us for a workshop, where this lesson was only a small part of the day. But we have also given short talks about this topic.

The last lecture before school holidays is always a special one. On one hand, the students are tired from the semester. On the other hand, they look forward to the school-free time and they are excited. Hence, it is a special challenge for the teacher to choose a topic which is interesting enough to elate the students in this situation.

In January 2011, we performed the presented lesson at a high school in Zurich in the framework of a regular informatics course. The lesson was for a programming class (in 10th school year, ages 17 to 18) and it was the last lecture before holidays. We wanted to give the students some insights to an interesting and important topic. Altogether we had 53 students.

A question which arose from time to time during the semester was, why is informatics important for everyday life. The goal was to show with this sample lesson the importance of informatics using cryptology as an example.

We aimed to create a lesson which elates all students, even those who are not very interested in informatics.

4.1   Evaluation

Influencing or changing students' attitudes to school subjects is generally a very hard objective, even when teaching classes over a very long time. Since attitudes are difficult to observe or to quantify, gaining evidence of a change in the way students think about a specific matter is a complex task, requiring formal empirical tests and deep know-how in cognitive psychology. The short survey we are going to present and to discuss aims to help confirming or rejecting the impression we had during the lecture. It clearly does not fulfil the requirements of a formal empirical study.

We were interested, whether this short introductory lesson into cryptology has an effect on the opinion of the students. Therefore we conducted a post-questionnaire four weeks after this introductory lesson took place. 53 students had to rate six statements with a rating between "total agreement" (rated 5) to "total disagreement" (rated 1). This questionnaire is not meant to be a full evaluative test, but merely a way to get some feedback (see Fig. 2).

The first question covered the previous knowledge about the topic. On average, this question was rated with a 2.5, which is higher than we expected. This could indicate that the topic is interesting enough, so that the students had already read something about it. A questionnaire from a comparable group of students showed that 50 % of them did not know the meaning of the expression "cryptology". In that group we had 87 students. One possible explanation of the relatively high value in this study is that the students were familiar with the term "cryptology".

With the second statement, we wanted to know whether we did a good job on getting the attention of the students. With an average rating of 4, the students attested that the instruction was indeed interesting, none of them answered that it wasn't interesting at all and 80 % rated the statement higher than 3. Certainly we hope that we did a good job, but we also believe that this in part means that cryptology is indeed an interesting subject.

The third point stated: "I would like to know more about the topic". 70 % answered that they agree or totally agree with this statement. Only six students do not want to know more about the topic. As this question was also answered very positively, we hope that the student's interest is strong enough that they might conduct some further reading on their own.

With statement 4, we wanted to find out whether the students believe that cryptology is indeed used in real life rather than just being a theoretical toy. Although the average of 3.8 is quite high, we think we could do better here. The quite similar statement 5 was on the other hand answered with much more agreement (average 4.3). It is good to see that the students approve that cryptology is important for today's communication. Possibly the previous question was rated lower because none of the students has ever used cryptology outside of school.

We are surprised with the outcome of the last statement. We hoped that the students would agree much more with this proposition. However, the question is formulated very vague. If the statement would have been "Everyone should know that cryptosystems exist and can protect your data" for example, the answers would probably look quite different.

The most important conclusion we made is, that after a few weeks, the students still remember this lecture. This topic is strongly related to their real life and therefore, this lecture seemed to influence the students attitudes towards informatics. Although, this lecture was meant to be only a short overview, some knowledge transfer happened.

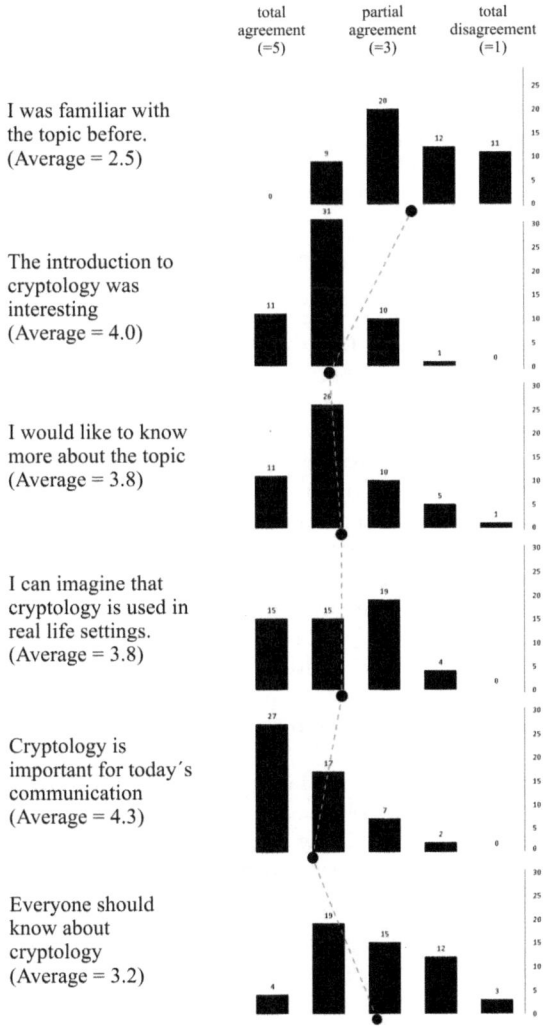

Fig. 2. Chart showing the graphically prepared data

5   Conclusion

In this paper, we introduced a classical cryptology lesson in which the students take the position of the cryptanalyst. The teacher takes the role of the code designer and presents the students some cryptosystems, and the students try to break these systems. Students are able to do that and this gives them a feeling of success. Also, the interesting history about cryptology motivates the students for further cryptology lessons.

This sample lesson can of course also take place in a math class since cryptology is a topic strongly related to both mathematics and informatics.

References

1. Bauer, F.L.: Decrypted Secrets: Methods and Maxims of Cryptology, 4th edn. Springer, Heidelberg (2006)
2. Bell, T., Fellows, M., Witten, I.H.: Computer Science Unplugged - Off-line activities and games for all ages (1999), http://www.csunplugged.org
3. Beutelspacher, A.: Cryptology. Mathematical Association of America, Washington, DC (1994)
4. Bruss, D., Erdélyi, G., Meyer, T., Riege, T., Rothe, J.: Quantum cryptography: A survey. ACM Comput. Surv. 39 (July 2007)
5. Freiermuth, K., Hromkovic, J., Steffen, B.: Creating and testing textbooks for secondary schools. In: Mittermeir, R.T., Sysło, M.M. (eds.) ISSEP 2008. LNCS, vol. 5090, pp. 216–228. Springer, Heidelberg (2008)
6. Freiermuth, K., Hromkovič, J., Keller, L., Steffen, B.: Einführung in die Kryptologie. Vieweg+Teubner (2009)
7. Hromkovič, J.: Sieben Wunder der Informatik. Vieweg+Teubner (2008)
8. Hromkovič, J.: Algorithmic Adventures. Springer, Berlin (2009)
9. Keller, L., Komm, D., Serafini, G., Sprock, A., Steffen, B.: Teaching public-key cryptography in school. In: Hromkovič, J., Královič, R., Vahrenhold, J. (eds.) ISSEP 2010. LNCS, vol. 5941, pp. 112–123. Springer, Heidelberg (2010)
10. Kerckhoffs, A.: La cryptographie militaire. Journal des sciences militaires IX, 5–38 (1883)
11. Kerckhoffs, A.: La cryptographie militaire. Journal des sciences militaires IX, 161–191 (1883)
12. Koblitz, N.: Cryptography as a teaching tool. CRYPTOLOGIA: Cryptologia 21(4), 317–326 (1997)
13. Singh, S.: The Code Book. Doubleday (1999)
14. Wätjen, D.: Kryptographie. Springer, Heidelberg (2008)

Little Beaver – A New Bebras Contest Category for Children Aged 8–9

Monika Tomcsányiová and Peter Tomcsányi

Department of Informatics Education,
Faculty of Mathematics, Physics and Informatics, Comenius University,
842 48 Bratislava, Slovakia
{Monika.Tomcsanyiova,Peter.Tomcsanyi}@fmph.uniba.sk

Abstract. The teaching of Informatics in lower and upper secondary education has a long tradition in Slovakia. In 2009, a new subject, Elementary Informatics ("Informatická výchova" in the Slovak language), was introduced into primary education. This paper describes its aims and relates how contests can be a form of learning and a way to test children's knowledge. Secondary schools have the Bebras Contest, and in this article we present the process of transforming tasks from the international database of Bebras tasks so that they are appropriate for the national contest in Slovakia. Furthermore, in order to facilitate participation of primary school children in the contest, we prepared and tested tasks in a new contest category, Little Beaver. The paper presents criteria of an appropriate contest task for children who are between 8 and 9 years old. Several tasks, which were solved by them in the pilot year of the contest, are included.

Keywords: Informatics, primary education, contest, Bebras Contest, children at the age of 8-9.

1 Introduction

Slovakia has a long tradition of teaching informatics. It is taught in the lower and upper secondary schools as a subject of general education. To support this subject, several textbooks were published (see [1]), which stress the understanding of informatics as a separate subject and not only as part of other subjects in which ICT are used.

Informatics is taught in all the grades and is divided into 5 topics:

- Information Around Us
- Using ICT for Communication
- Workflows, Problem Solving, Algorithmic Thinking
- Principles of ICT
- Information Society

Its content and range at each level of education is designed with respect to the abilities and needs of the students at their respective ages. The school reform of 2008 introduced a new subject to primary education, Elementary Informatics. This new subject includes the same topics as mentioned above.

I. Kalaš and R.T. Mittermeir (Eds.): ISSEP 2011, LNCS 7013, pp. 201–212, 2011.

2 The Subject Elementary Informatics

In the school year 2009/10, the new subject of Elementary Informatics for primary schools was established. It is intended for children from grades 2 to 4, who are 7, 8 and 9 years old at the beginning of their respective grades. The aim of this subject at this educational level is to guide children to use digital technologies effectively during searching and processing information, when solving problems and during general learning activities. Elementary Informatics also prepares children to understand the basic terms and mechanisms important to solving various kinds of problems by using digital technologies. It teaches children how to use the Internet as a communication tool, for learning, for information researching and for the presentation of information. Another important aim of the subject is to give the pupils the basics of algorithmic thinking and the ability to solve problems by using digital technologies.

The standard for the new subject, (see [2]), defines the abilities and skills of the pupils very generally. Therefore the choice of activities depends very much on the teacher. At this time, a textbook is available only for the 2nd grade (see [3]). There is no official textbook for the 3rd and 4th grade. However, we believe that a creative teacher is quite able to prepare many different tasks, problems or inter-subject projects that will help his/her pupils to gain skills in using digital technologies. We realise that for a teacher who has had no specialised education in informatics as a science, it will be a tough task, so in order to help them, we implemented a project of additional education in informatics for teachers in Slovakia.

Another way of educating both teachers and pupils is a contest. It can be an in-school activity or out-of-school activity, which helps pupils to enhance their abilities and skills in a given area, (see [4]). Therefore we decided to implement a new category for primary schools in the successful Bebras contest.

3 The Bebras Contest

The first Bebras contest was organized in Lithuania in 2004. The main aim was to promote interest in Information and Communication Technologies (ICT) to all school students, and to encourage children to use modern technologies more intensively and creatively in their learning activities.

Several other countries started to organize national Bebras contests, (see [5]), and in the school year 2007/2008 Slovakia joined the contest as well (see [6]). Since then the number of Slovakian participants has increased each year. In the school year 2010/11 there were 22139 contestants from ages 10 to 18, in four categories: Benjamins, Cadets, Juniors and Seniors[1].

For this kind of contest it is very important to create interesting contest tasks that will attract many students. Another role of the tasks is to differentiate students according to their understanding of informatics terms, how they understand different ways to represent data, and whether they can use their knowledge effectively and creatively in solving problems.

---

[1] In Slovakia Benjamins are aged 10–12, Cadets 13–14, Juniors 15–16, and Seniors 17–18. In other countries the ages may vary.

3.1 The Tasks

Task proposals are designed by each country's representatives and then are submitted to the international database. The best ones are selected by the participants of the International Bebras Task Workshop (see [7]). This collaboration among countries is possible because many of them have included informatics into the curricula of lower and upper secondary education (even though the content may differ from country to country). In order to use a task in a national contest, the participating country has to re-evaluate tasks from the international database, and select the tasks which match their curriculum. If the country decides to use a task from the international database, it has to perform the steps described below.

- **Translate the task** – The tasks in the international database are in English. Therefore, they must be translated into the language of the country.

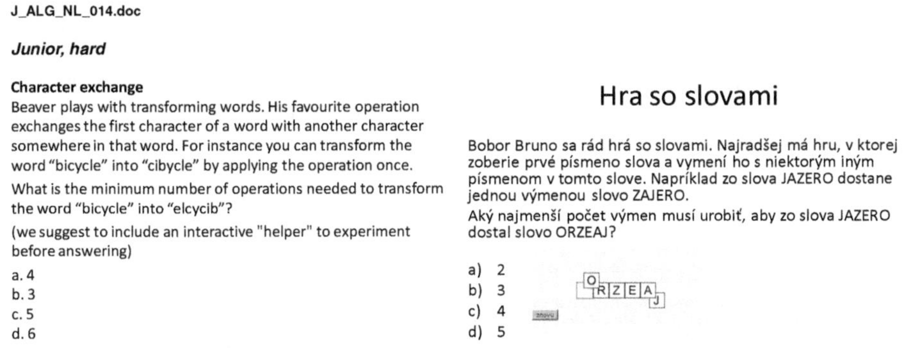

J_ALG_NL_014.doc

Junior, hard

Character exchange
Beaver plays with transforming words. His favourite operation exchanges the first character of a word with another character somewhere in that word. For instance you can transform the word "bicycle" into "cibycle" by applying the operation once.

What is the minimum number of operations needed to transform the word "bicycle" into "elcycib"?

(we suggest to include an interactive "helper" to experiment before answering)

a. 4
b. 3
c. 5
d. 6

Hra so slovami

Bobor Bruno sa rád hrá so slovami. Najradšej má hru, v ktorej zoberie prvé písmeno slova a vymení ho s niektorým iným písmenom v tomto slove. Napríklad zo slova JAZERO dostane jednou výmenou slovo ZAJERO.

Aký najmenší počet výmen musí urobiť, aby zo slova JAZERO dostal slovo ORZEAJ?

a) 2
b) 3
c) 4
d) 5

Fig. 1. The process of translation. The left picture shows the original English, the right one shows the final Slovak task with an interactive helper, category Junior 2008/09, easy task.

- **Adapt and translate the data** in the task – Many tasks contain data which are either in the accompanying images or directly in the assignment, e.g.: names, terms, menus of actual computer programs, etc. Each country has to replace them so they correspond to its tradition and terminology.
- **Specify the category** in the national contest – In each participating country, particular knowledge, skills or abilities may be learnt or achieved in a different grade or at a different age, so the task could be moved to a different category than the one suggested in the international database.
- **Reword and sometimes simplify the task assignment** – The task should not contain overly long texts. Correct formulations must be found which take into account the translated language and the chosen category.
- **Redraw the images** or draw new ones – Original images in the database are often schematic or they do not exist at all or they do not correspond to the reworded task.
- **Implement the software** – Interactive tasks need to be implemented as small applets. It depends on the manner in which that particular country implements interactive tasks. This often differs from the software programmed by the country that designed the task.

The fastest way

You want to visit your friend as quickly as possible. You drive from your house to your friend's house. The numbers on the map express how many minutes the car takes to drive each part of the way.

Mark the fastest path clicking on the parts of the way.

The map

Tom found a map leading to a treasure. There are written numbers in several boxes and he remembers that at the time of drawing it he wrote in the boxes the shortest distance from the box with the treasure. Secondly he remembers that you can walk from any box to the next one only by their adjacent side.

Mark the box with the treasure.

For example: here is a picture of the map and you can see the written numbers and the marked steps leading towards the red treasure.

Fig. 2. Sample interactive tasks showing the Slovakian interactive application, Find the Path, Senior, 2009/10, medium, The Map, Benjamins, 2010/11, hard

The contestants in a particular country solve the adapted tasks using the specific contest system of the country. In some countries contestants solve tasks online (e.g., in Slovakia); in other national contests the schools are given a computer program in which students solve the tasks offline. Nevertheless in each participating country the students solve the tasks directly at the computer.

Domino

John makes up a rectangle of domino tiles. A tile is attached to the previous one in the line if their number of dots equal. The picture shows what he has already built. He has seven extra tiles which are located in the bottom of the picture.

Try to finish the rectangle by dragging the tiles into the three empty boxes. If you doble-click a tile, it will turn.

In how many different ways can the rectangle be completed?
a)1
b)2
c)3
d)4

Houses

You need to colour some houses by using the **fill** tool and two colors - red and blue. They must be painted in such a way that no two of them are equal. Each house must have both its triangle and square part painted.

a) 2
b) 3
c) 4
d) 5

You can try it out by clicking on the squares and triangles below.

How many different houses can be painted?

Fig. 3. Sample tasks with interactive helpers. Domino, Juniors, 2009/10, hard. The little houses, Little Beaver-pilot, 2010/11, hard.

The advantage of this approach over using pen and pencil is not only the use of ICT and developing computer literacy, but it also adds the possibility of solving **interactive tasks**. Students solve these by manipulating objects within the task assignment (dragging the objects, clicking on them) and leave the object(s) in a state that corresponds to the task's assignment.

Another type of task which involves interactivity, both to make the solution easier and help to achieve the correct understanding of the assignment, is a multiple choice task with an **interactive helper**. Similarly to an interactive task, the interactive helper provides direct manipulation of the task's objects in order to experiment with them. According to [8], experimenting is a problem-solving scheme. It helps to better understand the task and transform it into the contestant's mind. After experimenting, and according to their own experience with the task, the contestants can enter the solution by ticking the correct choice just like in a multiple choice task (see figure 3).

We think that for 8- and 9-year-olds the aforementioned two kinds of tasks that include interactivity should be preferred over purely multiple choice tasks.

4   Preparing the Little Beaver

Since Elementary Informatics started in 2009, the first pupils who went through it will be in 4th grade in November 2011. Therefore, for the school year 2011/2012 we decided to create a new Bebras contest category called **Little Beaver**.

While preparing the tasks we assumed that pupils will have gained the knowledge and skills needed for solving them in their **Elementary Informatics lessons**. However, we also assumed that pupils use digital technologies **at home**, too. Experiments show (see [9]) that children who are 8 and 9 use their own electronic equipment (a mobile phone is used by 95.7 % of pupils) or the equipment found in their home, i.e. television, camera, computer, and printer. Using and controlling these devices becomes natural for them and they gain the needed skills very quickly just by using this equipment. They do mention the standardization of the devices' controls and layouts and the use of icons and texts for labelling controls have the same or similar meaning on different devices. Therefore, children who are 8 and 9 know the icons for turning the devices on/off, and the icons for running or stopping the video, and so on. Of course, pupils expect the same controls on other devices and in PC programs or PC games.

4.1   Differences between Bebras and Little Beaver

When preparing tasks for children who are 8 and 9, we must consider the differences between them and pupils aged 10 to 18 for whom the other categories of the Bebras contest are intended. First of all, their psychological and psychomotorical development [10] completely differs from students aged 10 to 18.

Following the principles for designing software for children (see [11]) we will focus on the following aspects:

- Pupils of these ages **read more slowly** and do not always understand the text correctly – We will try to reduce the text, make the task assignment succinct and make the font large enough.
- Children at this age need to work with concrete objects within the software and they do not understand abstraction – the task assignments will be created with concrete people, things and events that they can experience in real life or meet in fairy tales.
- The objects' size and distances on the computer screen have to be adapted to the child's age – we will try using graphics which appeal to children, similar to illustrations in story books.
- Children at this age cannot focus on a task for a long time – we will shorten the duration of the contest to 30 minutes and the number of tasks to 12 (for the older categories we use 15 tasks and 40 minutes).

Besides the above-mentioned differences, in the category of Little Beaver, the majority of tasks will be either interactive or tasks with an interactive helper. According to [10], manipulation of objects is very important for children at this age, more important than it is in higher categories.

4.2 Using a Bebras Task for the Little Beaver

In this section we will present two tasks included in the Benjamins category in previous years. While preparing the new contest category, we consider how and whether to use this type of task. We also make clear whether it is possible and appropriate to put them in the new contest category. We select the tasks with the knowledge and skills learnt when they studied Elementary Informatics from the beginning of the school year 2009/10 till the day of the contest.

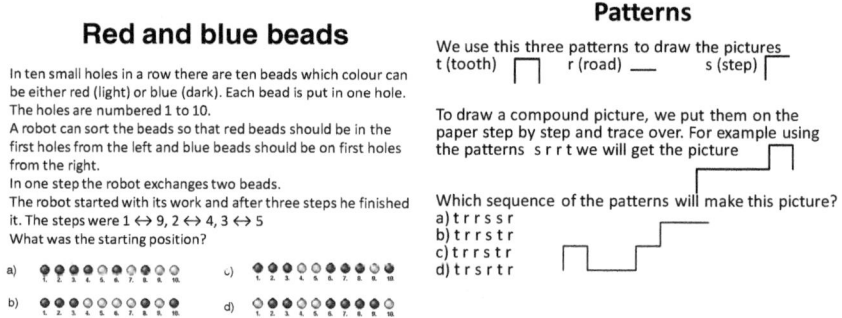

Fig. 4. Red (light) and blue (dark) beads - Benjamins, 2009/10, medium, Patterns - Benjamins, 2010/11, easy

If we decide to use this type of task in the new category, Little Beaver, we will consider:

- **task classification** into one of the topics of Elementary Informatics, (see [4]), and its categorization according to the international agreement (see [12]),
- **task adequacy to** the children's age in this contest category; here we consider the task's motivation and the graphic used,
- **task formulation** to be ready for transformation and translation for national contests in other countries,
- classification according to its **difficulty**: easy, medium, hard; the task difficulty is often determined by its formulation and interactivity,
- an **interactive** variant of the task – whether it is possible to make the task interactive or prepare an interactive helper for it; we prefer this type of task in the Little Beaver category,
- in the case of non-interactive tasks we consider the appropriate **distractors** which allow us to detect misconceptions in children's knowledge when evaluating their answers,
- **reasoning** for the right solution of the task – sometimes pupils do not understand why other answers are not right even when the correct answer is told to them immediately after the contest.

5   From the Idea Till Its Implementation

Next sections present how a creative author can prepare a task for the new contest category. We will show how to find good topics and how to convert the first idea of a task into its final implementation.

5.1   The Idea

When looking for an appropriate contest task one can be inspired in different ways. Informatics background and concepts of informatics can be found in real problems that children at the age of 8 and 9 could experience. One of the author's inexhaustible sources can be the Internet. The author can use either websites designed directly for children or any other sites which are realistic enough for children of the given age.

Just imagine that you and your children want to go to the cinema and together you look at the cinema's web page. It contains the reservation system of the cinema which allows selection and reservation of seats and even allows buying tickets for them. We were actually using this site with our children and we started to think about how to use it in the Little Beaver contest. We asked ourselves:

- Could we use this idea in a contest task for children who are 8 and 9?
- What is the informatics background of the task?
- Which of the children's knowledge, abilities and skills will be tested? What activity must the visitor and the children do on the web site to reserve the tickets?
- Does a child of 8 or 9 have the knowledge and abilities to manage the task?
- Is the problem realistic enough for children of this age? Do they understand it?
- Are we able to prepare an adequate and appropriate task out of this idea?

If one decides to use this idea to create the task, he/she must concretize it further for our assignment:

- How large should the cinema be to be appropriate for the Little Beaver Category?
- How should visualize the occupied seats be visualized? Should they be in colour and abstract or is it better to locate figures or faces on them?
- How will the task be controlled? Will the child click on the seats or will he place figures on the seats?

5.2   Choosing a Formulation

At first we design several types of tasks (without their concrete formulation) by using the idea of seat reservation. Later we decide which type of task is the most appropriate for the contest. We formulate the chosen task as accurately and clearly as possible. For example, we consider the following:

- In an empty cinema, we search for several possibilities for seat placement: seats which are in the middle of the row, which are as far from the screen as possible, as near to the screen as possible, about in the middle of the row, at the edge of a row, close to the cinema's exit.

- In a partly occupied cinema, we search for three places which are not too far from the screen, only the seats at the edge, and the seats close to the cinema's exit etc.
- In a quite occupied cinema, we want to find several seats next to each other – for this assignment we can prepare several rooms with various numbers of seats to search for in various occupied cinemas.
- We also consider preparing a task that has no solution. If we use it in the contest, it will be significantly more difficult than a task which can be solved.

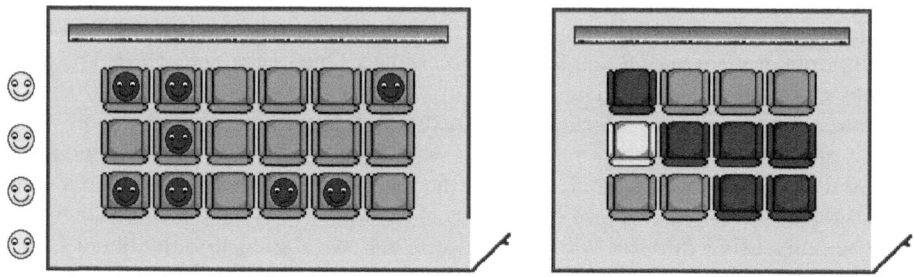

Fig. 5. Solving the task by placing the faces (left) and finding the solution by more abstract way (right)

5.3 The Implementation

To finalize the task we must:

- determine the cinema's **size and shape,** determine how many seats will be required to solve the task – if the cinema is too large and has too many seats to search for, it can discourage the children from solving the task at all,
- design the exact task assignment - the difficulty of the solution will depend on it,
- decide the manner of manipulating the objects in the task, which will determine how the children will find the solution. In this case of an interactive type of task there are two possibilities: children will drag the faces (see Fig 5 left) or we can represent the occupied seats by colours. The latter will make reserving seats more abstract since only the colour of the seat will change (Fig 5 right). For the Little Beaver Category, it would be more natural to solve the task by dragging the faces,
- after the concretization of the task, we define the difficulty of it – easy, medium, hard,
- draw the images which are interesting and eye-catching for the children,
- program the task so that it can be used in our national contest system or we can present it to teachers and the international committee who select tasks for the international database.

6 The Pilot-Run of Little Beaver

In order to try our ideas for the tasks and also to gain some knowledge of how difficult they would be for real children, we decided to run a pilot contest.

Since the decision was made quickly, we did not use our complete contest software for the new category. Instead we adapted our demo version of the contest for this purpose. Normally we use the demo version to allow anybody to enter the archive of tasks from previous years and try them out. Since it does not require any registration prior to the contest, it has been easier to adapt. The downsides of this decision are (1) that the same pupil can participate several times (i.e. he/she can try to solve the same task several times), and (2) that without adding extra measures (see later in this chapter), we cannot generate any lists of results per school.

The interest for the contest among the teachers has been unexpectedly high. We even had to stop registration when we exceeded the estimated number of 1000 participants.

We prepared 11 tasks, 4 easy, 4 medium and 3 hard ones. The 12th task was to write down the name of the participant. We added this task to be able to create a list of participants (and winners) for the participating teachers. This task has been excluded from evaluation so that no pupil would be forced to write down his/her name. 7 out of 11 tasks were interactive. Only 4 tasks required choosing the correct answer from a list of 4 answers, and one of those tasks had an interactive helper. Therefore, 8 out of 11 tasks involved some type of interaction and manipulation. Only the remaining 3 tasks had to be solved without any kind of on-screen experiments.

The evaluation scheme has been modified (compared to other Bebras categories) to avoid non-integral numbers of points. To avoid a negative total number of points we added a bonus of 24 points to each participant. Because the 12th task was not evaluated, the actual minimum of total points is 3 and the maximum is 87.

The pilot-run was executed on 29th of April 2011 between 10:00 and 13:00. By the end of the contest there were 1370 participants from about 60 schools who finished the contest and another 220 participants who started the contest but did not finish.

We decided that, for the purpose of analysis, we would use only data from those participants who identified themselves and who either finished the contest or answered more than half of the tasks. These rules were satisfied by 1216 contestants.

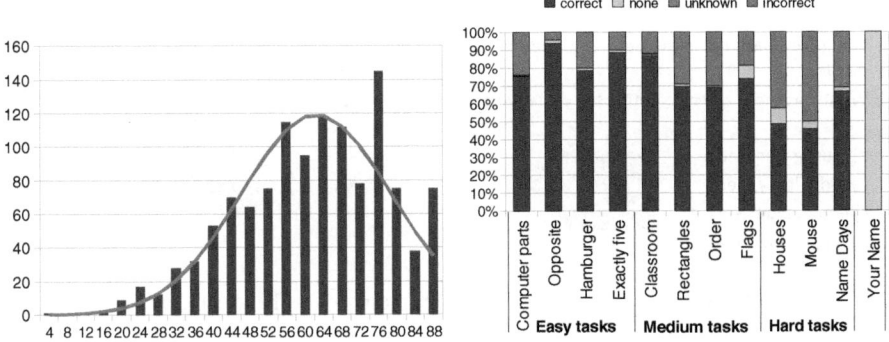

Fig. 6. Left: distribution of achieved total points (bars) compared to the theoretical normal distribution computed from the mean and standard deviation of the data set (line). Right: the rate of correct, incorrect, absent and unknown (due to a software problem on client's side) solutions for each of the 12 tasks (note that there are very little "unknown" solutions to see them in the small graph). The 12th task has not been evaluated and therefore it has been treated as 100% not answered.

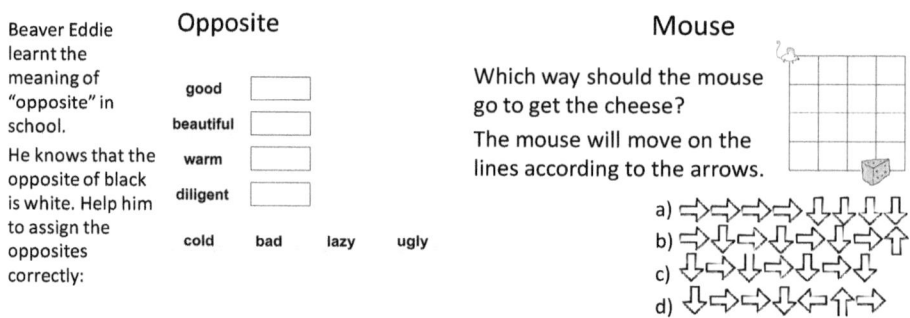

Fig. 7. The easiest task (Opposite) and the hardest one (Mouse)

We draw these conclusions from the diagrams of figure 6:

- The tasks were quite easy for the children; the hardest was solved correctly by 46% of the participants and the easiest was solved correctly by 94% of the participants.
- We were quite successful in predicting the complexity of the tasks (unusually successful compared to the other categories of our Bebras Contest). The three tasks that we estimated as being hard were actually the three hardest ones in the contest. And only two tasks stand out of their respective group: the task Computer Parts (figure 8) was estimated as easy, while it is a medium task according to the results, and Classroom (not shown in this paper) was estimated as medium while it turned out to be easy.
- The extra high peak at the values between 73 and 76 points can be explained by the fact that this interval includes the value 75 points, which could be achieved by incorrectly solving only one hard task, which happened quite often.

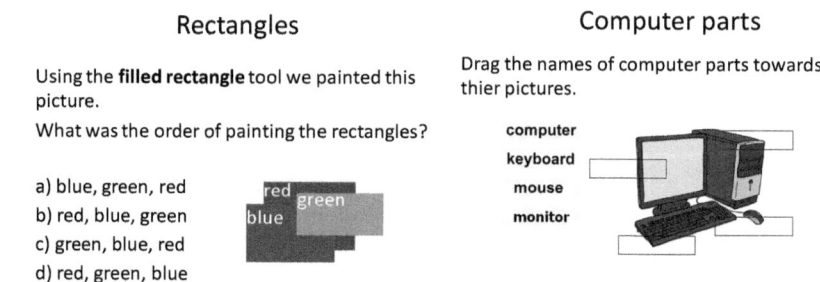

Fig. 8. Two more tasks with interesting findings (see the text for details)

When we analyzed more closely the incorrect answers, we have found some interesting facts:

- The most frequent incorrect answer to the task Mouse (figure 7) was answer a), (36% of participants chose it). The answer contains a sequence of moves to the left followed by a sequence of down moves. Our initial guess was that many participants had not been able to understand the more complicated programs in the other answers at all. Then a teacher complained that the picture of the cheese has been connected only to the right gridline, not to the top and right one. So the children assumed that there is no path leading to the cheese from its left side or top side. Therefore they tried to approach it from the right using the answer a).

- The most frequent incorrect answer to the task Houses (figure 3 at the end of chapter 3) was 2 (22%). This means that many children did not consider painting the roof of the house with a different colour than its body.
- The most frequent incorrect answer to the task Rectangles (figure 8) was c) (20%), which gives just the opposite order of how the rectangles were really painted. This is a bit surprising because we expected that the vast majority of children would have already painted rectangles in a graphical editor.
- The task Computer Parts (figure 8) proved to be harder than we expected. Actually 24% of participants reversed the labels for "computer" and "monitor". Even though we see this mistake among young children quite often (they often react to the command "switch off the computer" by switching off the monitor), we did not expect such a high ratio of incorrect answers. Maybe the word "monitor" is too hard for some of them and we should have considered the word "screen" instead.

We have received several positive reactions from the teachers. They reported that the children enjoyed the contest and many of them considered the tasks easy even if they did not have a high score.

Some teachers reported technical problems. We must analyse them in more detail before the regular round of the contest in November 2011. Some of the problems may arise from the fact that many primary teachers are less experienced in setting up the computers and that many computer labs are set up by external administrators.

The pilot-run was successful. It proved that the tasks were appropriate for the considered age group (even possibly too easy) and that most of the involved primary teachers were able to organize the contest successfully in their schools.

7   Conclusion

In our article we presented the development of our idea for a new category in the Bebras contest, which we call Little Beavers. We explained the history and reasons for its creation, the way we select tasks and the procedure we used to execute a pilot-run of the contest as well as the results of that pilot.

We offer these thoughts and experiences to anyone who would like to implement the same ideas in his/her own country.

References

1. Kalas, I., Winczer, M.: Informatics as a Contribution to the Modern Constructivist Education. In: Mittermeir, R.T., Sysło, M.M. (eds.) ISSEP 2008. LNCS, vol. 5090, pp. 229–240. Springer, Heidelberg (2008)
2. Štátny vzdelávací program Informatická výchova, Príloha ISCED 1 (The standards for Elementary Informatics in Slovakian), http://www.statpedu.sk/

3. Blaho, A., Salanci, Ľ., Chalachánová, M., Gabajová, Ľ.: Informatická výchova pre 2. ročník, aitec, Bratislava (2010) (Textbook of Elementary Informatics, in Slovakian)
4. Kalas, I., Tomcsanyiova, M.: Students' Attitude to Programming in Modern Informatics. In: 9th WCCE: IFIP World Conference on Computers in Education – Education and Technology for a Better World, pp. 127–135. Porto Alegre (2009)
5. International Web page of the Bebras contest, http://www.bebras.org/
6. Hrušecká, A., Pekárová, J., Tomcsányi, P., Tomcsányiová, M.: Informatický bobor – nová súťaž v informačných technológiách pre žiakov základných a stredných škôl. In: Proceedings Didinfo, Banská Bystrica (2008) (Informatic Beaver – a new contest in ICT for lower and upper secondary students, in Slovakian)
7. Dagiene, V., Futschek, G.: Bebras International Contest on Informatics and Computer Literacy: Criteria for Good Tasks. In: Mittermeir, R.T., Sysło, M.M. (eds.) ISSEP 2008. LNCS, vol. 5090, pp. 19–30. Springer, Heidelberg (2008)
8. Polya, G.: How to Solve It: Summary. Princeton University Press, Princeton (1957)
9. Gregussová, M., Kovacikova, D.: Sú naše deti vo virtuálnom prostredí v bezpečí? (Are our children safe in the virtual environment? In Slovakian), http://www.zodpovedne.sk/index.php
10. Piaget, J., Inhelder, B.: The Psychology of the Child. Basic Books, New York (1969)
11. Chiasson, S., Gutwin, C.: Design Principles for Children's Technology, http://hci.usask.ca/publications/2005/HCI_TR_2005_02_Design.pdf
12. Futschek, G., Dagiene, V.: A Contest on Informatics and Computer Fluency Attracts School Students to Learn Basic Technology Concepts. In: 9th WCCE IFIP World Conference on Computers in Education – Education and Technology for a Better World, pp. 1–9. Porto Alegre (2009)

What's the Fun in Informatics?
Working to Capture Children and Teachers
into the Pleasure of Computing⋆

Violetta Lonati[1], Mattia Monga[2], Anna Morpurgo[1], and Mauro Torelli[1]

[1] Università degli Studi di Milano,
Dipartimento di Scienze dell'Informazione
[2] Università degli Studi di Milano, Dipartimento di Informatica e Comunicazione,
Via Comelico 39, 20135 Milano, Italy
{violetta.lonati,mattia.monga,anna.morpurgo,mauro.torelli}@unimi.it

Abstract. The importance of computer science education in secondary, and even primary school, has been pointed out by many authors. But too often pupils only experience ICT, both at home and at school, and confuse it with computer science. We organized a game-contest, the *Kangourou of Informatics*, with the aim to attract all pupils (not only the talented ones), expose them to the scientific aspects of informatics in a fun way, and convey a correct conception of the discipline. Peculiarities of the game are its focus on team work and on engaging pupils in discovering what lays behind what they experience every day.

Keywords: informatics and education, learning contests.

1 Introduction

"*I think that it's extraordinarily important that we in computer science keep fun in computing. When it started out, it was an awful lot of fun.*" The pioneer Alan Perlis (1922–1990) stated this very clearly as a priority goal, but today informatics is rarely depicted as a rewarding activity in itself. Although our lives are inevitably interwoven with computers, software, and automatic computations, most of the general public perceives informatics as technological overhead to be avoided as much as possible. In fact, informatics is great in producing ready-to-use abstractions that can be used as black boxes for specific tasks, but the abstract nature of computation is there, and it needs to be discovered to open up real innovation and human development. So we are increasingly facing the problem of convincing non-experts that the desirability of informatics is not just in its value as an instrument, but also as a human challenge, both intellectual and social. A basic understanding of the underlying principles of informatics should therefore be common knowledge among the general public as are the principles of other more traditional sciences. However, not everyone is convinced

---

⋆ This work was partially funded by Google under the Computer Science for High Schools (CS4HS) program.

I. Kalaš and R.T. Mittermeir (Eds.): ISSEP 2011, LNCS 7013, pp. 213–224, 2011.

that teaching computing to kids is desirable or, perhaps, *feasible*. Indeed, informatics could play a very important role in education as it promotes the ability to deal with abstract concepts, the acquisition of the scientific method, and a pragmatic approach to problem solving and work organization [11,2]. Historically, as it became evident that the computer was beginning to play a very important role in our every day life (in 1982 "The Computer" was named Machine of the Year by Time Magazine), projects to introduce informatics in schools were developed. Informatics education started with programming, and the Logo programming language was used for a while for that aim. But as the personal computer made its way through in schools and homes, the attention shifted from informatics to the use of the PC and its applications: computers were sometimes bought by schools before developing informatics curricula and preparing teachers adequately. Most was left at the individual initiative and fantasy. The effect was a serious misconception of what informatics is about, and informatics is still nowadays too often confused for ICT [4]. Thus, even if the last decades have seen an increasing exposure of children to computer applications, the basics of information, computing, and software sciences are still not part of the curricula of primary and secondary schools in several parts of the world. According to a report of ACM and the Computer Science Teachers Associations (`http://www.acm.org/runningonempty`) the number of computer science courses in the US has decreased in the past five years, and when the schools offer a course, it is usually an elective one. Moreover, most of these courses are in fact about information technology literacy rather than a look-ahead toward the issues introduced by the automatic elaboration of information. In Italy the context is even worse: only computer literacy, mostly presented as an instrument to increase one's proficiency in other subjects, is indicated as an important educational goal, and even that is never taught as a specific subject by experts of the field (except for vocational, technical, schools). In most Eastern European countries informatics as a separate subject was established 30 years ago: training of programming skills used to get more attention at the beginning, while nowadays much more consideration is devoted to developing ICT skills [6].

1.1   The Many Facets of Informatics

Indeed basic computer application literacy is certainly a useful skill to acquire, but not at the cost of confusing pieces of knowledge that are largely independent: fluency in an application domain and familiarity with the computing disciplines.

Mirolo [13] points out that the term *informatics* is used in different contests with different meanings:

1. informatics as a science, providing its own peculiar key to interpret reality and its specific approach to problem solving;
2. informatics as a technology, concerning the characteristics, structure and working principles of the now ubiquitous hardware and software devices;
3. informatics as an instrument, providing practical tools to manage information in many different contexts.

It is not easy to give a precise and complete definition of what computer science is. Hromkovič [11] points out that computer science is for some aspects a meta science such as mathematics, as it investigates general categories; a natural science for other aspects, as it studies objects and processes, it deals with quantitative rules of natural processes, it investigates what is possible and what is impossible, and it uses the scientific method; it is a problem oriented and practical engineering discipline dealing with technical and management issues. It is therefore a challenging task to work for a correct perception of informatics. To clear the confusion from the ground up, we coined a specific word for the use of computer applications that does not require computing skills: we called that *applimatics* in [12]. Applimatics is not only the use of office automation tools, but also the use of a dynamic geometry program to explore Euclidean axioms. Even the use of things like the Alice [5] programming environment to explore the properties of bodies immersed in a 3D space and graphic animation techniques or the meritorious Logo are not, per se, informatics, if the emphasis shifts from computational thinking to geometry[1]. We liked the challenge and started the *Kangourou of Informatics* game-contest (http://kangourou.dsi.unimi.it), which aims at giving a correct view of computer science to both pupils and teachers, by exposing them to the scientific aspects of informatics, not considered in syllabi of most schools. The intention is to convey the correct conception of informatics in a fun way and attract all pupils, not only the talented ones. We identified, however, three main obstacles that have to be overcome to effectively propose computing games to children and (non expert) teachers. On the one hand there is the abstract nature of computing (mostly shared with mathematics), that makes it difficult to show the concepts by referring to physical objects and the fact that often specialized language and terms are needed; the games should avoid too abstract tasks and the use of jargon to propose the challenges (we sometimes propose jargon as a challenge in itself!). On the other hand, however, it is very important to avoid to completely disappoint the expectations of pupils about informatics; the link with applimatics to which they are daily exposed should be made clear as much as possible, to grow a consistent and fertile global picture.

In the following sections we will describe the main ideas that drove us in setting our game-contest: in Section 2 we briefly survey informatics competitions; in Section 3 we describe our experience in organizing the *Kangourou of Informatics*; in Section 4 we report the feedback we got from the participants; and in Section 5 we draw some conclusions.

2 Promoting Informatics through Competitions

Consistently with the idea that computing is fun when you engage in it, games and contests are rather common among computer science experts. The most famous contest is probably the International Olympiad in Informatics (IOI) [17,16]. The first IOI was held in Bulgaria in 1989 under the sponsorship of UNESCO. IOI is open to high school pupils throughout the world. The competition tasks

---

[1] Indeed Logo was originally proposed to improve the teaching of mathematics [15].

proposed to participants are of algorithmic nature, and, while they aim at stimulating interest in informatics and information technology, the explicit goal of IOI is to give recognition to young pupils from around the world who are the most talented in computing. On similar lines the ACM organizes an International Collegiate Programming Contest (ICPC) in which teams are given 5 hours to solve between 8 and 12 programming problems: the winner is the team which correctly solves most problems. Moreover, contests are increasingly common in the information security arena. In the '90s the DEF CON conference popularized a computer security wargame called "Capture The Flag" (CTF): each team is given a (virtual) machine or a small virtual network to defend from the attempts of intrusion of the other players. Teams are scored for both successful defense and attacks. CTFs evolved into full educational exercises to give participants experience in securing complex systems. In 2004 an international, academic CTF was started by the University of California at Santa Barbara, and several other similar contests are organized around the world and often attract online contestants from several countries and different educational backgrounds. These competitions, however, assume that participants have already acquired some expert skills in the field: an accompanying training is sometimes planned to bring the contestants to the level required to compete. IOI, for example, promotes local training sessions (often several weeks long) to teach high school pupils the programming skills needed for the Olympics. Thus, such competitions mostly address an audience that is already inside the fun of informatics or had at least the opportunity to see it from a technical point of view. This chance occurs either because of a meeting with an especially motivated teacher or, more likely, to pupils attending a specialized educational program. Thus, most pupils are out of reach of these contests, at least in countries (like Italy) where computing is not explicitly part of generic curricula: they often end up in knowing only the applimatics face of informatics, but the challenges of information technology are rarely presented to them.

An important step in popularizing the fun of informatics to a public with little knowledge of the technicalities of the discipline is the rise of non-specialized game-contests. In 2009 we started the *Kangourou of Informatics* by piggy-backing the experience of our University in organizing the *Kangourou of Mathematics*. In Kangourous the games have the explicit intent to attract the maximum number of pupils without aiming at any national selection nor at a comparison between countries and we embraced exactly those goals to foster the knowledge of computing to an audience as vast as possible. In fact, we recently discovered that Kangourous were inspiring several other people long before us. In 2004 Valentina Dagiene started a Kangourou-inspired game-contest in Lithuania under the name of *Beaver*[7,9,8,1]. The game became international in 2007 and it has several points of contact with our proposal: indeed we would really love to join our efforts. In the following sections we try to describe the peculiarities of our approach in organizing our game-contest and what we want to convey to pupils about our discipline.

3   The Kangourou of Informatics

A game-contest, the *Kangourou des Mathématiques*, was created in 1991 in France by André Deledicq on the model of the Australian Mathematics Competition, with the goal of contributing to the popularization and the promotion of mathematics among young people. The success was immediate also thanks to the associated distribution of a massive and pleasant documentation on mathematics to the participating pupils and their teachers. The French experience was exported abroad, first to Europe and then to other continents through an international association, *Kangourou sans frontières*, founded in France in 1995. The association's aim is to promote the spreading of a basic mathematical culture by all means and, in particular, by organizing the annual game-contest to be held on the same day in all participating countries. The game now counts millions of participants among primary and secondary school kids (48,000 in Italy in 2010). In Italy, which joined the association in 1999, the game is organized in cooperation with the Mathematics Department of the Università degli Studi di Milano. As a consequence of the effectiveness of the event, in 2005, at the yearly Kangourou Sans Frontières international meeting, the Romanians suggested to extend the game-contest to foreign languages and informatics. In 2008, Kangourou Italy invited two Informatics Departments of Università degli Studi di Milano, AICA, and SDA-Bocconi to study a formula for an informatics game-contest. On the basis of our previous experiences with IOI (as trainers for the Italian national IOI team) and UCSB International Capture the Flag (our team won in 2007 and and was placed among the first positions in 2008 and 2009), we took up the challenge. The first contest was held in 2009, and the third edition this year.

3.1   Organization of the Game-Contest

The *Kangourou of Informatics* is a team game-contest and is held yearly nationwide. Each team is composed of four pupils; at the moment there are two categories: category "medie" for junior high school pupils (age 11–13) and category "biennio" for pupils in the first two years of high school (age 14–15), with slightly diversified difficulty levels. We plan to extend the game also to younger pupils. There are two phases:

1. a qualifying round carried out on-line and organized locally in March in the schools under the supervision of the schools' teachers;
2. a final round in May for the best 24 teams, held in Mirabilandia, an amusement park near Ravenna.

In both rounds the problems are presented in a playful way, since one of the main goals is to make pupils enjoy themselves while discovering what informatics is. The game is mostly skill-oriented, as no prior knowledge can be assumed. Moreover, pupils are allowed and encouraged to use the Web in both rounds to search for information or hints they may need to answer the questions. The problems are at different levels of difficulty and spanning various aspects of informatics, from logic to programming, from grammars to concurrency.

The questions are chosen so that it is very unlikely to score zero points, but also very difficult to totalize the maximum score. The game-contest is partly self-supported with the subscription fees and has commercial sponsors which also provide some prizes for the first three teams in each category and their schools. After the qualifying round a booklet on the contest is published and sent to the schools. It contains the problems as presented during the game and, for each problem, the answer and how it can be obtained and the informatics topic to which the problem refers. Both the booklet and the on-line test are available on the Kangourou's website. The booklet should, respectively could, help teachers discover this discipline and overcome their fear of inadequacy regarding computer science (we refer to mathematics and technical education teachers, since in many Italian schools computer science is not taught as a separate subject). Besides it is intended as an aid in preparing for the final round and for the next year's competition both for pupils and for teachers.

3.2   Qualifying Round

The qualifying round is carried out on-line locally by the schools under the supervision of the teachers. The software and the problems are downloaded from a central server. The teams have to solve around 10 problems (see below for examples) in about an hour (it may vary slightly from year to year). The software keeps the time, collects the answers and sends them to the server for evaluation. We think the use of the computer in the contest, though not essential for the questions per se, is very important in order to keep the link explicit with what pupils and teachers normally associate with the word informatics. We believe it is very important to clinch to the fact we are not discussing something different from what the world expects from computer science experts: we are just trying to explain better what is under the hoods.

3.3   Examples of Qualifying Challenges

This subsection describes some of the games we proposed.

Maze. One of the problems was how to reach a treasure in a maze. The setting was a sequence of rooms, each one identified by a number, n, and with two exit doors, the left one leading to room number $2n$ and the right one leading to room number $2n + 1$. The goal was to reach room number 69, where a treasure was kept. The task was to guess the correct sequence of doors that had to be passed through to reach the target room. If an erroneous door was opened, the player lost one of his or her three lives and had to start over. The task could be solved with different approaches: by trial and error, by drawing the graph of the rooms' connections and finding there the path connecting the two rooms, by noticing that each room in this maze can be reached only from one other room and working backwards from room 69 to room 1: all even numbered rooms are reached by a left door, all odd numbered rooms by a right door. 20% of the teams (16% in the "medie") got the maximum score (no lives were consumed in unsuccessful tries) and another 31% (same percentage in the "medie") were able to solve it with two or three trials.

A pipe and filter game. The problem stated that Riccardo has two lists of soccer teams: OldTeams and NewTeams. In the former he collects the teams of the 2010 edition of Champions League and in the latter the ones participating in the 2011 edition. He wants to know which teams participated for the first time, *i.e.*, which teams are in NewTeams but not in OldTeams. In order to compute the solution, Riccardo may combine three programs:

1. catenate, which is able to append a list to a given one;
2. dups, which returns a list of duplicates in a list;
3. uniqs, which returns a list of unique values in a list.

The game was proposed in two different versions for "medie" and "biennio". The easier one ("medie") asked to identify a correct solution among different sequences of program applications, expressed in natural language: the correct solution was "Catenate NewTeams to OldTeams; find duplicates in the result; catenate them to NewTeams; find unique elements in the result". The quiz was answered correctly by 40% of the teams. The "biennio" version asked to build a solution by combining the programs: each program was represented by a graphic block with changeable inputs and an arrow denoting the output as input *filtered* by the block. The goal was to produce a sequence by connecting a block to the preceding one: in fact all the blocks but the first had one input fixed to the output of the previous as in a pipeline of filters; the number of blocks was limited to a maximum of five. Although the pupils were driven by the graphical scaffolding, this version turned out to be much more difficult. Very few teams were able to find an optimal solution with only three blocks, and some others found a solution with 4 or 5 blocks (in total, a correct solution was built by the 19% of the teams only).

Secret sentences. In this problem a group of friends decided to defend their hiding place using secret sentences. To enter, one has to pronounce a valid sentence, which is built according to the following rules:

- valid sentence: a simple sentence or a simple sentence followed by an adverb, a verb, and a valid sentence;
- simple sentence: an article followed by a description of a mouse, or an article followed by a description of a cat;
- description of a mouse: "mouse", or a description of a mouse followed by "white";
- description of a cat: "cat", or a description of a cat followed by "fat", or "red" followed by a description of a cat.

The pupils had to identify which sentences from a given set were valid and allowed entrance into the hidden place. The topic dealt with by the question is the use of a formal grammar to generate valid sentences. The grammar is based on the pupils' natural language, but allows some counter-intuitive constructions: an adjective may follow a noun, and the same adjective may be repeated consecutively, thus the pupils had to be careful not to confuse the concept of correctness under the rules of the game and correctness in their mother language. The question (proposed only for the "biennio" category) turned out to be quite difficult: only 15% of the teams solved the game perfectly and gained the full score. However, in total 92% of the teams were able to get some correct answers.

3.4 Final Round

The final round of the *Kangourou of Informatics* game-contest takes place in Mirabilandia amusement park and is reserved to 24 of the best qualified teams (at most one team from each school is admitted, in order to discourage cheating in the qualifying round). Participating in the final round is seen as a prize for all finalists: they are hosted with no extra expenses by Kangourou Italia and have a free admission ticket to all park attractions during the days of the finals.

The round is organized as follows. In the first evening a welcome meeting with all team members and accompanying teachers is scheduled, where they get information about regulations and the timetable. On the next morning all finalists are called for the actual contest that lasts the whole morning. During the contest, accompanying teachers are offered some lessons and labs, that suggest topics and innovative ways to convey informatics culture to pupils. The rest of the day is occupied with the grading process which is usually quite time-consuming. The day after the contest, a conference is organized for both finalists and accompanying teachers, which ends in the award ceremony.

The final competition consists of several games of different types. First, each team member is involved in a game aimed at obtaining its account to access a computer and an envelope containing the other assignments. These can be tackled in any order and have to be handed in all together at the end of the contest. Such assignments usually consist of written problems and questions, tasks to be developed on the computer, search tasks on the Web.

With the aim of stressing some peculiar aspects of informatics, we impose some rules that promote team work and smart use of limited resources: there is only one computer for every two teams and each session expires after 15 minutes. Moreover, computers are located at some distance from the place where team members are seated. Hence, it is important to manage some sort of team organization and time expenditure. The best teams often show an interesting subdivision of roles (the "programmer", the "logical thinker", the "pony express", the "checker", . . .), some multi-threading of tasks (*i.e.,* combination of individual work, work in pairs, and discussions in the whole team), and a good ability in avoiding downtime (e.g., when the computer is unavoidable, time is not wasted by waiting, but it is used to establish which search should be done later on the Web, to design the solutions to be implemented, and to prepare instructions or entire programs to edit and execute later on the computer).

The initial game actually starts the evening before the contest, just after the welcome meeting: each team receives some material to read or to examine, which can be used to get some training for the competition. All team members get immediately excited by this unexpected preview and plunge into the material trying to catch it and figure out how it could be used! For instance, in the contest of 2009, we divided each team into 2 subgroups, gave the team user-name to one subgroup and the team password to the other one, and asked them to exchange their information on a public channel using the cryptographic method described the day before, in order to complete the whole picture about their own accounts. In 2010, instead, the game was based on a riddle about trees, leaves, and heaps. This year, the game required to understand a non-losing strategy, described in a visual way, for tic-tac-toe. This part of the contest does not increase the score of the teams, but gives them the possibility to gain time over the other contestants. If a team does not succeed in solving the game within a fixed amount of time, the account information is revealed, a penalty is assigned, and the team is given the next assignments.

Written problems are quite standard, in that they focus on typical problem-solving issues in the field of data representation, cryptography, combinatorics, algorithms and data structures, languages (grammars and automata), games, logic, and so on. However it should be emphasized that problems are not in multiple-choice form. Usually open questions are asked and a brief explanation that motivates the answer is requested. The score assigned to this part of the test obviously takes into consideration the correctness of the answer, but also depends on other parameters: is the motivation relevant, accurate, complete, well-written?

The tasks requiring the use of computers usually concern some aspects of informatics we consider typical and educational.

Problem-solving and programming. We assume no prior knowledge about programming language syntax. Hence we propose visual programming environments than can be understood and mastered easily and quickly, like Logo [15], Scratch [14], Etoys [10]. The goals to achieve are also graphical: contestants had to build paths around obstacles in 2009, to draw colored geometrical pictures in 2010, to find a way of escape from mazes this year. A sequence of exercises is proposed, with increasing difficulty: first only sequential instructions are needed, then loops, sub-routines, and variables become necessary as the level increases. Some bonus points can be gained when a solution is correct with respect to several instances, in order to introduce the idea of generality of algorithms.

Text description and structuring. Teenagers are familiar with word processors that follow the *what-you-see-is-what-you-get* paradigm, like Microsoft Word or Microsoft PowerPoint. However, they do not have any idea about how these programs work internally, and have no model about how a text is represented by such kinds of software suites [3]. Moreover, they usually use these programs naively and often ignore the possibility to structure texts with *paragraph styles*, or to distinguish semantic structure from typographic aspects. This somehow corresponds to a deeper, logical, confusion between the meaning of the content and its appearance. During the final round, we always schedule a game that aims at discovering these aspects of text processing, asking to provide some written "formal description" of a structured text. In order to keep abstraction and concreteness in touch, we specifically designed and implemented web-based programs that were at disposal of the teams to test their solutions: such programs use these formal descriptions (also if partial) to re-build the document. Thus pupils have the possibility to check what seems right and what is wrong, and they have the chance to correct mistakes, with a trial-and-error approach.

In 2009 we proposed the use of a *wiki*, *i.e.* the use of a simplified mark-up language that uses, for instance, quotation marks for emphasized text, marker "-" and proper indentation for itemized lists, and so on. The teams received a written tutorial about basic formatting rules with some examples and were asked to reproduce a full formatted text. In 2010 the game was titled "Into PowerPoint" and it asked for the description of a slide presentation in a tabular form. Contestants received a pdf with the slide presentation and a printed page with the set of all text elements occurring in the presentation. They first had to classify each slide according to a list of possible page formats. Then they had to fill in a table having a row for each text element, by defining the slide containing it and its position in the slide. Finally they had to describe some graphical aspects (text color, background color, thickness, alignment) of some basic elements of the presentation (titles, footer, ...)[2]. In 2011 we played with HTML: here the main goal was to recognize the tree structure of a given text, and mark each inner node according to its structural role (title, section title, paragraph, link, ...).

Informatics jargon. Web searches, which are allowed for all proofs, are mainly used for this part of the contest. Contestants have the chance to discover the meaning of words or expressions they already heard but whose meaning they ignore (often also because they are not translated into Italian), to dissolve false ideas, and to correct the improper use of some technical expressions. Moreover, the intent is to tell some stories or anecdotes about the history of informatics, computer scientists, and informatics practitioners.

In 2009 the teams had to fill in the blanks in some short stories. Missing keywords were for instance Alan Turing, artificial intelligence, CamelCase, spam, or Linux. In 2010 we used a bad translator from English to Italian to produce some incomprehensible and funny texts about files, hardware components, or operations on computers. We exploited the double meaning of words – for instance, *file* is also the English word denoting "a metal tool with a rough surface for cutting or shaping hard substances or for making them smooth"; or the word *folder* can be seen as formed by the verb "to fold" plus a suffix "-er", and hence can be translated with the meaning of a "person who bends something, especially paper or cloth, so that one part lies on top of another part"! In 2011 we joked with geek humour, inspired by those T-shirt with geeky sentences written on front: "to understand recursion first one must understand recursion", "there are only 10 types of people in the world: those who understand binary and those who don't", "there's no place like 127.0.0.1", "2B | [^B]{2}". Each sentence had to be related to an image: a fractal picture for recursion, an image representing the conversion of a decimal number into binary, a warmy home for the localhost IP,

---

[2] It should be noticed that the tool implemented to support teams in testing their solution actually built the presentation using LATEX-Beamer instead of PowerPoint since this gives much more control on structure and styles!

a portrait of Shakespeare for the "to be or not to be" regular expression. Clearly, to guess the right matchings, teams had to first understand the general meaning of the sentences or, at least, their context.

4   Participants Feedback

We are currently setting up collaborations with teachers to be able to collect a formal feedback from the participants and correlate it with the participants' performance and backgrounds. We already monitored some of the web pages and the student newspapers the pupils wrote after the participation. They describe the experience as fun and challenging. A common remark is that some of the issues, for instance those about jargon, are quite strange and unexpected; probably their comments are due to the fact that they did not associate with computer science many of the terms and expressions we proposed (like *regular expressions* or *camel case* or *root*). The choice of organizing the game around teams has a positive side effect: in order to be able to participate, pupils more focused on computers solicit friends with lesser interest, and the teams are indeed heterogeneously composed.

A strong positive feedback also comes from the teachers. In particular, they welcome the availability of the booklets we prepare: whenever they want to discuss informatics in their classes, they usually have to choose between specialized literature (mostly out of reach for a high school audience in a non technical environment) and ICT/business-oriented publications. The descriptions of the Kangourou challenges, instead, provide a way to introduce a topic and how it relates to applimatics and ICT and the references provide pointers to further studies. We are also considering the preparation of other types of support that are more specific to class work.

5   Conclusions

We firmly believe that informatics is a scientific discipline with an important educational value and sufficiently basic to be taught as a fundamental formative subject. It is also our belief that computing is indeed fun and that, by playing and working in team, pupils can discover some of the most important aspects of this discipline. We used these characteristics to organize the *Kangourou of Informatics*, a two rounds national game-contest. In the qualifying round we can attract a wide audience to the basic issues of informatics. In the final round we can engage pupils in more challenging tasks and offer refresher courses and laboratories to their teachers. In this way, pupils can experience what informatics is, and go beyond ICT, which is the usual approach of schools to this subject.

References

1. Antonitsch, P.K., Grossmann, A., Micheuz, P.: Beaver, Kangaroo and classroom situations: A promising symbiosis. In: Hromkovič, J., Královič, R., Vahrenhold, J. (eds.) ISSEP 2010. LNCS, vol. 5941, pp. 16–31. Springer, Heidelberg (2010), http://www.issep2010.org/proceedings_of_short_communications.pdf

2. Barr, J., Cooper, S., Goldweber, M., Walker, H.: What everyone needs to know about computation. In: Lewandowski, G., Wolfman, S.A., Cortina, T.J., Walker, E.L. (eds.) Proc. of the 41st ACM Technical Symposium on Computer Science Education, SIGCSE 2010, Milwaukee, Wisconsin, USA, pp. 127–128. ACM, New York (2010)
3. Ben-Ari, M.: Constructivism in computer science education. In: Lewis, J., Prey, J., Joyce, D., Impagliazzo, J. (eds.) Proc. of the 29th SIGCSE Technical Symposium on Computer Science Education, Atlanta, Georgia, USA, pp. 257–261. ACM, New York (1998), http://doi.acm.org/10.1145/273133.274308
4. Brinda, T., Puhlmann, H., Schulte, C.: Bridging ICT and CS: educational standards for computer science in lower secondary education. In: Brézillon, P., Russell, I., Labat, J.M. (eds.) Proc. of the 14th Annual SIGCSE Conference on Innovation and Technology in Computer Science Education, ITiCSE 2009, Paris, France, pp. 288–292. ACM, New York (2009), http://doi.acm.org/10.1145/1562877.1562965
5. Carnegie Mellon University: Alice website, http://www.alice.org
6. Dagiene, V.: Teaching information technology in general education: Challenges and perspectives. In: Mittermeir, R.T. (ed.) ISSEP 2005. LNCS, vol. 3422, pp. 53–64. Springer, Heidelberg (2005), http://dx.doi.org/10.1007/978-3-540-69924-8_27
7. Dagiene, V.: Information technology contests – Introduction to computer science in an attractive way. Informatics in Education 5(1), 37–46 (2006), http://www.mii.lt/informatics_in_education/htm/INFE069.htm
8. Dagiene, V.: Sustaining informatics education by contests. In: Hromkovič, J., Královič, R., Vahrenhold, J. (eds.) ISSEP 2010. LNCS, vol. 5941, pp. 1–12. Springer, Heidelberg (2010), http://dx.doi.org/10.1007/978-3-642-11376-5_1
9. Dagiene, V., Futschek, G.: Bebras international contest on informatics and computer literacy: Criteria for good tasks. In: Mittermeir, R.T., Sysło, M.M. (eds.) ISSEP 2008. LNCS, vol. 5090, pp. 19–30. Springer, Heidelberg (2008), http://dx.doi.org/10.1007/978-3-540-69924-8_2
10. Squeak Etoys website, http://www.squeakland.org/
11. Hromkovic, J.: Contributing to general education by teaching informatics. In: Mittermeir, R.T. (ed.) ISSEP 2006. LNCS, vol. 4226, pp. 25–37. Springer, Heidelberg (2006), http://dx.doi.org/10.1007/11915355_3
12. Lissoni, A., Lonati, V., Monga, M., Morpurgo, A., Torelli, M.: Working for a leap in the general perception of computing. In: Cortesi, A., Luccio, F. (eds.) Proceedings of Informatics Education Europe III. IFIP, pp. 134–139. ACM, New York (2008), http://www.dsi.unive.it/IEEIII/atti/PROCEEDINGS_IEEIII08.pdf
13. Mirolo, C.: Quale informatica nella scuola (2003) (in Italian), http://nid.dimi.uniud.it/pages/materials/discussion/educazione.pdf
14. MIT Media Lab: Scratch website, http://scratch.mit.edu/
15. Papert, S.: Teaching children to be mathematicians vs. teaching about mathematics. Artificial Intelligence Memo AIM-249, MIT (1971)
16. Verhoeff, T.: The role of competitions in education (1997), http://www.win.tue.nl/~wstomv/publications/competit.pdf, presented at Future World: Educating for the 21st Century, a conference and exhibition at IOI 1997
17. Verhoeff, T.: The IOI is (not) a science olympiad. Informatics in Education 5(1), 147–159 (2006), http://www.mii.lt/informatics_in_education/htm/INFE078.htm

Criteria for Writing Exams Which Reflect the K12 CS Foundations Study Material

Haim Averbuch[1], Tamar Benaya[2], and Ela Zur[2]

[1] Alon High School,
101 Oshiskin St., Ramat Hasharon, Israel
haimav@zahav.net.il
[2] The Open University of Israel, Computer Science Department,
108 Ravutzky st., Raanana, Israel 43107
{ela,tamar}@openu.ac.il

Abstract. All high school students in the Israeli high school system are required to take matriculation exams in the main subjects studied in high school. The matriculation exams are similar to the American AP exams in that they are external nationwide exams. The high school teachers are required to prepare the high school students for the matriculation exams. This is done by using internal exams which should reflect the matriculation exams. The first part of the Israeli Computer Science high school curriculum includes the courses Foundations of Computer Science 1 and 2. In this paper we describe the structure and content of the matriculation exam of Foundations of Computer Science 1 and 2. We define several criteria for constructing the internal exams. In addition, we discuss the following issues: What qualities contribute to an effective exam? How well the internal exams reflect the matriculation exams? How teachers can prepare better exams?

Keywords: Computer Science Education, K-12, Assessment.

1 Introduction

In the last three decades, there has been considerable activity surrounding Computer Science (CS) curricula on all levels (beginning with ACM Curriculum Committee on Computer Science [1] through Computing Curricula 1991 [2] and up to Computing Curricula 2001 [3]). Notable is the high-school curriculum designed by the special ACM task force [4], and in particular the new K-12 curriculum [5]. The goal of the new K-12 curriculum (2nd edition), was to create a 4-level curriculum that could be widely disseminated, accessible to every high school student in the US. Its aim was to enable every CS student to understand the nature of the field and the place of CS in the modern world. Students need to understand that CS combines theoretical principles and application skills. They need to be capable of algorithmic thinking and of solving problems in other subject areas and other areas of their lives.

In Israeli high schools, every student must study at least one subject in depth, in addition to general studies which include Mathematics, English, History, Literature etc. The highest level of studies is the 5-point (as opposed to 3- or 4-point) program,

I. Kalaš and R.T. Mittermeir (Eds.): ISSEP 2011, LNCS 7013, pp. 225–235, 2011.

each point representing 90 class hours. CS is one of the subjects that high school students can select to study in depth.

The Israeli CS high school curriculum was designed in the early 1990s and first implemented in 1995. One particular principle underlying the curriculum is the interleaving of theoretical principles with application skills. This interleaving notion is specifically termed in the Israeli curriculum as the "zipper approach" [6]. The curriculum has two versions, a 3- and a 5-point version. The 3-point program includes two mandatory core units, Foundations of Computer Science 1 and 2 (denoted by FCS1 and FCS2), which present the foundations of algorithmic thinking and programming. The 5-point program is intended for more advanced students. It includes the 3-point version and the fourth mandatory unit, Software Design, which is an extension of FCS1 and FCS2. The third and the fifth units can be chosen from several alternatives. A detailed description of the program is given in [6, 7]. The curriculum has been updated to some extent since it was first implemented, shifting, for example, from procedural to Object Oriented languages, and is now undergoing a more extensive modification.

All high school students are required to take matriculation exams in the main subjects studied in high school. The Israeli matriculation exams are similar to the American AP exams in that they are external nation wide exams. The high school teachers are required to prepare the high school students for the matriculation exams. This is done by using internal exams which should reflect the matriculation exams. The final grade in the subject tested is calculated as the average of the matriculation exam and an internal grade which is partially based on the internal exam and the student performance throughout the year. It is important that teachers will be familiar with the matriculation exams in order to prepare their students in the best way possible [8, 9]. The questions in the matriculation exams, like the AP exams, should test the intended concepts accurately, unambiguously, and without bias [10].

In this paper we describe the structure and content of the matriculation exam of FCS1 and FCS2. We define nine criteria for constructing the internal exams. In addition, we discuss the following issues: What qualities contribute to an effective exam? How well the internal exams reflect the matriculation exams? How teachers can better prepare their students for the exams?

The next section describes the courses FCS1 and FCS2 which are the first part of the high school CS curriculum.

2   Foundations of Computer Science

2.1   Syllabus

The courses FCS1 and FCS2 are taught in Java or C#, according to the CS teachers' preference in each school. No preliminary knowledge is assumed. Each unit is designed for 90 school hours (3 weekly hours). According to the recommendations of the Israeli CS curriculum committee, FCS1 is intended for the 10th grade, and FCS2 – for the 11th grade; though in many schools both units are taught in the 10th or 11th grade.

The topics of the two course units "zip" conceptual notions with actual algorithmic (computer program) structures. In addition, they embed both algorithm design notions and object oriented elements. The former is presented in both units, whereas the latter is primarily displayed in the 2nd unit.

FCS1 includes the basic computational elements of: variables, conditional execution, and repetition, interleaved with the conceptual notions of: design stages, correctness, and efficiency. FCS2 interleaves the essential OOP notions of objects and classes with the mutual algorithmic/OOP notion of patterns, and the general notion of problem solving [11].

2.2 Matriculation Exams

The content of the matriculation exam of FCS1 and FCS2 is supposed to reflect the material described above.

The duration of the exam is three hours. The exam is divided into three sections according to the Bloom taxonomy [12] which is demonstrated by Thomson et al. in their paper [13].

The first section contains 5 mandatory ten point questions which test basic skills such as knowledge and comprehension.

The second section includes 3 fifteen point questions out of which the students are required to answer two. The questions in this section are application questions which require the students to solve problems to new situations by applying acquired knowledge. The questions in this section may require writing a small program, or writing a sub-program and demonstrating its use or tracing a given program. This section requires the use of sequential and/or nested patterns.

The third section includes 2 twenty point questions from which the students are required to answer one. The questions in this section require analytic and synthesis skills. This section requires writing a complete program which includes: defining appropriate sub-tasks, defining main variables and data structures and implementing the code including documentation.

3 Analysis of the Internal Exam

We conducted a preliminary study in order to evaluate the internal exams and to determine how well they reflect the matriculation exam in FCS1 and FCS2. We first describe the criteria according to which we analyzed the internal exams and then we discuss and analyze a sample of internal exams.

We defined the following nine criteria which we believe the teacher should abide by while constructing the internal exams. The criteria are divided into two groups: the first group relates to an overview of the exam as a whole and the second group deals with elements within single questions.

Overview of the exam
- *Criterion 1: Exam structure.* The exam structure should be identical to the structure of the matriculation exam.

- *Criterion 2: Balanced coverage of the material.* The exam should include a variety of data types and structures, a variety of patterns, etc.
- *Criterion 3: Balanced mixture of types of questions* such as: designing, writing, tracing, detecting identical solution, etc.
- *Criterion 4: Appropriate difficulty level of the exam* which can be solved within the given time constraints.
- *Criterion 5: Equal difficulty of all questions in sections* where given a choice.

Elements within single questions:

- *Criterion 6: Clear phrasing.* Each question should be phrased clearly and accurately with no ambiguity.
- *Criterion 7: Representative examples.* Algorithmic questions should include several representative examples of input values and their expected output.
- *Criterion 8: Error free questions.*
- *Criterion 9: Refraining from nonsense questions*, such as tracing a segment of code which does not present a solution to a meaningful algorithmic problem.

We analyzed thirty one FCS1 and FCS2 internal exams composed by different teachers and schools from the years 2006-2009. All the exams were written by teachers who taught FCS1 and FCS2 for at least five years. We analyzed these exams according to the criteria presented above. We present below an analysis of the different criteria followed by relevant examples from this sample.

3.1   Overview of the Exam

Criterion 1: Exam structure

We found that all of our sample exams exhibited an identical structure to the structure of the matriculation exam.

Criterion 2: Balanced coverage of the material

Tew and Guzdiel identified concepts that a wide variety of introductory CS courses had in common [14]. Similarly, we defined the following elements which should be included in the internal exams:

- Data types such as: simple variables (integer, double, boolean, character, etc.), arrays (one and two dimensional), strings and classes.
- Control structures such as: sequential, conditional and repetitive.
- Algorithmic patterns, such as: counting, accumulation, maximum, average, decomposing a number into its digits, searching, sorting, merging, etc.

Some of the questions should include a combination of sequential and nested structures and algorithmic patterns described above. For example, finding the maximum prime number in a given series of numbers combines nested structures and patterns.

We found that 85% of our sample exams exhibited a well-balanced coverage of the material.

Criterion 3: Balanced mixture of types of questions

Programming literacy includes capability of both reading programs and writing programs. Lister et al. claim that code comprehension and tracing ability are prerequisite skills to problem solving [15]. We believe that the exam should have a balanced mixture of types of questions such as:

- Designing a problem solution;
- Writing a program/algorithm or section of code;
- Tracing a section of code. In general, in the case of trace questions we recommend to also ask the student to explain the purpose of the algorithm;
- Comparing solutions in order to detect their identity;
- Providing representative input values for given outputs;
- Checking correctness of an algorithm or program;
- Filling in missing sections of code;
- Detecting the purpose of the algorithm or program;
- Defining the signature of a method;
- Adapting a given solution to a problem constraint.

The first section of the exam has 5 questions which should exhibit the variety mentioned above. Unfortunately, not all exams abide by this criterion. We found an exam that includes 4 tracing questions out of 5. Another exam included two trace questions and three programming questions. On the other hand, we found several exams with a good mixture of types of questions. For example, an exam which included the following types of questions: trace, program correctness, programming a section of code, programming a complete program, combination of correctness and comparison.

The second section of the exam requires answering 2 questions out of 3. These questions are application question. The questions in this section may require writing a small program, writing a sub-program or tracing a given program. These questions should have several sections exhibiting a mixture of question types described above. For example, we found a question which included several question types. The question presented a method followed by sections requiring the students to trace the execution, to explain the purpose of the method, to give representative input values, and to provide an equivalent more efficient solution.

The third section of the exam requires analytic and synthesis skills. The students must select 1 question out of 2 and it is usually a larger design and programming question which includes: defining appropriate sub-tasks, defining main variables and data structures and implementing the code including documentation.

The following question is an example of a good question for the third section of the exam because it is a relatively complex question. The question requires the ability to decompose the problem into sub-problems and to apply a combination of sequential and nested structures and algorithmic patterns and to use of a variety of data structures.

A square matrix is called "max in rows" if its maximal value appears in more than one row. For example, the square matrix presented below is "max in rows" because its maximal value 7 appears in two separate rows (in the 1st and 3rd row).

Develop and write an algorithm which inputs a 20x20 integer matrix. The algorithm will check if the matrix is a "max in rows" matrix. If it is, the algorithm should print all the row numbers in which the max appeared, otherwise, the algorithm should print a statement indicating that the matrix is not a "max in row" matrix.

4	5	-9	7
5	3	0	4
1	7	4	3
6	0	1	2

a.   *Divide the problem into sub-tasks (at least 2). Define the goal of each sub-task and write its pre-conditions and post-conditions.*
b.   *Select the main variables, define their data types and write their roles.*
c.   *Write a program which implements the developed algorithm.*

Criterion 4: Appropriate difficulty level of the exam which can be solved within the given time constraints

The duration of the matriculation exam is three hours and therefore the internal exam should be written accordingly. As mentioned above, the exam has three sections. The first section which has five ten point questions should be solved in one hour. We claim that all the questions should be of equal difficulty level and should be solved in approximately ten minutes. From looking at our sample exams, we found some exams that included questions of different levels of difficulty. For example, the following question appeared in one of the exam:

a.   *Develop an algorithm which receives in variables N and M two positive integers. The algorithm will display the product of all the numbers between N and M which can be divided by 7 without a remainder.*
b.   *Given an array C of a length 10:*

8	2	-2	3	7	1	5	9	6	-4

Create a new array B of length 10 according to the following rule: B[I] receives C[9-I] – 1.
What will be the value in B[7]?

The problem with this question is that it is composed of two unrelated sections and therefore more difficult to solve within the given time constrains. We suggest that such a question should appear as two separate questions.

The second section of the exam requires 2 questions out of 3, each of which should be solved in half an hour. The following is an example of a good question for this section which we found in our sample exams:

The roller coaster safety regulation in an amusement park requires that a person's height will be at least 1.60 meters and his weight will be less than 100 kg in order to participate in the ride. The roller coaster has 80 seats and 150 persons are in the line. Write a program which inputs the height and weight of each person before he attempts to enter the roller coaster. The program performs the followings:

a. *Prints a message indicating whether the roller coaster was full when it departed?*
b. *Calculates and prints the number of people that were rejected and the number of people remaining in the line.*

This is a good question for this section because it requires the students to solve a problem to a new situation by applying acquired knowledge.

The third section of the exam requires 1 design and programming question out of 2 and should be solved in one hour. We found that most of the questions in the third section of our sample exams exhibited appropriate difficulty.

Criterion 5: Equal difficulty of all questions in sections where given a choice

Sections 2 and 3 of the exam are sections that include a choice. The questions in these sections should be of equal difficulty. We present two questions from our sample exams which appeared in section 2 of the same exam. These questions exhibit different levels of difficulty:

Question 1

The following section of code is required to define an array of 50 elements and to assign it the following values:

0.5 1 1.5 2 2.5 3 3.5........

```
double[] arr = new double[50];
for (int i=0; i < arr.length * 2; i++)
    arr[i] = i / 2;
```

Is this section of code correct?
If not – correct the mistakes so the required output is achieved.

Question 2

a. *Write a method that receives an array and returns the largest value in the array.*
b. *Elections for the class committee were carried out among 5 candidates. Write a program which inputs the students' votes (numbered 1 to 5). The input ends with a negative value. The output is the number of votes for each candidate and the id number(s) of candidate(s) which received the most votes. (Remark: use the method written in section a).*

The first question is too simple for this section and the second question is of appropriate difficulty level.

3.2 Elements within Single Questions

Criterion 6: Clear phrasing

The following question is an example of a question which is not clear:

A top model is defined by the following characteristics: height, age and weight measurement. Write an expression which evaluates to true if a model has a weight value less than 60 kg. and is at least 1.80 meters high or is older than 18.

It is unclear from the question which is the correct expression:

```
weight < 60 and  height > 180 or age > 18 or:
    weight < 60 and (height > 180 or age > 18)
```

Criterion 7: Representative examples

The following questions are examples of questions with representative input and output examples which are problematic as explained below:

Question 1

Nine-complement of a digit x is defined as 9-x. Nine-complement of a positive integer is a number where every digit is replaced by its nine-complement. Examples:

The nine-complement of 4 is 5, because 4+5=9
The nine-complement of 0 is 9, because 0+9=9
The nine-complement of 7 is 2, because 7+2=9
The nine-complement of 318 is 681
Write a method which receives a two digit positive integer and returns its nine-complement.

We found several problems with the representative examples: 1) The explanation of the examples is not directly derived from the definition; for instance, in the first example it should be 9-4=5 and not 4+5=9. 2) The examples are not balanced presenting three examples of one digit and one example of a three digit number and none of a two digit number.

Question 2

Given a one dimensional array of 55 integers, write a section of code which inputs an integer number (num) and outputs the number of pairs of sequential elements that their sum can be divided by num without a remainder.
For example: for the following array of size 8 and num=10

3	7	2	9	1	6	4	6

The output will be 3 because 3+7=10, 9+1=10, 6+4=10, 4+6=10.

We found two problems in the above question. The first is a mistake in the output which should be 4 and not 3. The second problem, which we want to emphasize, is that the sum of each pair in the representative example which divides by num is 10. We think that the representative example should include pairs whose sums are a variety of multiples of num.

Criterion 8: Error free questions

The following question is an example of a question which has an array index out of bounds mistake:

Given the following algorithm:

```
1. max ← a[p]
2. print max
3. for i from p+1 to n do:
      2.1  if a[i] > max
             2.1.1   max ← a[i]
             2.1.2 print max
```

a. Given p=3, n=8 and the following array a:

| 12 | 4 | 5 | 32 | 3 | 42 | 202 | 40 |

What will be the output?
b. Explain briefly what the algorithm does.

In addition, it is quite difficult to accurately explain what the algorithm does. Therefore, most students will not be able to gain the maximum of points for section b.

Criterion 9: Refraining from nonsense questions

The following question appeared in the first part of an internal exam. This question is suitable for the first part but it is an example of a nonsense question:

Given the following algorithm:

```
1.   number ← 25
2.   while number is greater than 0
      2.1If number is even
          2.1.1        number ← number/2
      2.2number = number-3
3.   print number
```

Use a trace-table to follow the above pseudo code and provide its expected output.

The pseudo code presented above does not present any meaningful algorithm and we believe that such algorithms should be avoided because they are useless. In addition, the algorithm should be more general by replacing the first line with an input statement instead of an assignment and then asking the student to trace the algorithm for a given input.

4 Summary and Recommendations

In this paper we provided nine criteria which we believe the teachers should abide by while constructing the internal FCS1 and FCS2 exams. We presented examples taken from a sample of internal exams written by experienced professional teachers.

We recommend that the teacher will prepare in advance an outline of the exam, delineating the data types, control structures and algorithmic patterns that the exam should include. After writing the exam, the teacher should ensure a balanced coverage of the material, by preparing a checklist of the different data types, control structures and algorithmic patterns. The teacher should go over the exam and mark the appropriate items on the list and make sure that the exam is balanced. Similarly, a checklist should be used in order to ensure that the exam presents a balanced mixture of question types.

Furthermore, the teacher should not only think of the exam solution but should actually solve the exam including running the programs. Such solutions insure that the exam's difficulty level is appropriate and that it can be solved within the given time constraint. Solving the exam also leads to error detection in the questions and further refinement of the exam.

We also recommend that the teachers give their exams to their peers to be solved by them. Unfortunately, many schools have only one CS teacher who works in isolation. This is due to the fact that a relatively small number of high school students select to expand their CS studies and this number is constantly decreasing. For these teachers we recommend the use of the Israel National Center for Computer Science Teachers Website [16] in order to interact with their colleagues. One of the main goals of the teachers' center is to promote pedagogical objectives, inspire colleagues and help them adjust to new courses and topics [17]. We recommend that the center will conduct workshops which will provide guidelines for teachers in exam preparation.

References

1. ACM Curriculum Committee on Computer Science, Curriculum '68 recommendations for academic programs in Computer Science, Comm. Assoc. Comput. Mach. 11 (1968)
2. Tucker, A., et al.: Computing Curricula 1991: A Summary of the ACM/IEEE-CS Joint curriculum Task Force Report. Comm. Assoc. Comput. Mach. 34, 69–84 (1991); Joint IEEE Computing Society/ACM Task Force on Computing Curricula, Computing Curricula 2001 Final Report,
   http://www.acm.org/education/curric_vols/cc2001.pdf
3. Merrit, S., et al.: ACM model high school computer science curriculum. Association for Computing Machinery, New York (1994)
4. Tucker, A., et al.: A model curriculum for K–12 Computer Science: Final report of the ACM K-12 Task Force Curriculum Committee, 2nd edn. (2003),
   http://csta.acm.org/Curriculum/sub/ACMK12CSModel.html
5. Gal-Ezer, J., et al.: A high-school program in computer science. Computer 28(10), 73–80 (1995)
6. Gal-Ezer, J., Harel, D.: Curriculum and course syllabi for a high-school computer science program. Computer Science Education 9(2), 114–147 (1999)
7. Drysdale, S., et al.: The year in review....Changes and lessons learned in the design and implementation of the AP CS exam in Java. In: Proc. of the 36th SIGCSE Technical Symposium on Computer Science Education, pp. 323–324 (2005)
8. Tymann, P.T., White, L.: The future of the AP CS program. In: Proc. of the 40th SIGCSE Technical Symposium on Computer Science Education, pp. 331–332 (2009)
9. Hunt, F., et al.: How to develop and grade an exam for 20,000 students (or maybe just 200 or 20). In: Proc. of the 33rd SIGCSE Technical Symposium on Computer Science Education, pp. 285–286 (2002)
10. Armoni, M., Benaya, T., Ginat, D., Zur, E.: Didactics of Introduction to Computer Science in High School. In: Hromkovič, J., Královič, R., Vahrenhold, J. (eds.) ISSEP 2010. LNCS, vol. 5941, pp. 36–48. Springer, Heidelberg (2010)

11. Bloom, B.S. (ed.): Taxonomy of Educational Objectives, the classification of educational goals – Handbook I: Cognitive Domain. McKay, New York (1956)
12. Thompson, E., Luxton-Reilly, A., Whalley, J., Hu, M., Robbins, P.: Bloom's Taxonomy for CS Assessment. In: Tenth Australasian Computing Education Conference (ACE 2008), vol. 78, pp. 155–161 (2008)
13. Tew, A.E., Guzdial, M.: Developing a Validated Assessment of Fundamental CS1 Concepts. In: Proc. of the 41th SIGCSE Technical Symposium on Computer Science Education, pp. 97–101 (2010)
14. Lister, R., et al.: A Multi-National Study of Reading and Tracing Skills in Novice Programmers. In: Working Group Reports from ITiCSE on Innovation and Technology in Computer Science Education, pp. 119–150 (2004)
15. The Israeli National Center for Computer Science Teachers' Website, http://cse.proj.ac.il
16. Israel National Center for Computer Science Teachers: "Machshava" – The Israeli National Center for High School Computer Science Teachers. In: Proceeding of the 7th SIGCSE Annual Conference on Innovation and Technology in Computer Science Education, Aarhus Denmark, vol. 234 (2002)

Author Index

GPSR Compliance

*The European Union's (EU) General Product Safety Regulation (GPSR)
is a set of rules that requires consumer products to be safe and our
obligations to ensure this.*

*If you have any concerns about our products, you can contact us on
ProductSafety@springernature.com*

In case Publisher is established outside the EU, the EU authorized
representative is:

Springer Nature Customer Service Center GmbH
Europaplatz 3
69115 Heidelberg, Germany

Batch number: 09474011

Printed by Printforce, the Netherlands